rhetorical climatology

rhetorical climatology

by a reading group

Chris Ingraham, John Ackerman, Jennifer Lin LeMesurier,
Bridie McGreavy, Candice Rai, and Nathan Stormer

MICHIGAN STATE UNIVERSITY PRESS | *East Lansing*

Michigan State University Press
East Lansing, Michigan 48823-5245

Library of Congress Cataloging-in-Publication Data
Names: Ingraham, Chris author.
Title: Rhetorical climatology : by a reading group / Chris Ingraham [and others].
Description: East Lansing : Michigan State University Press, [2023] | Includes bibliographical references.
Identifiers: LCCN 2023001798 | ISBN 9781611864793 (paperback) | ISBN 9781609177485 (PDF) |
ISBN 9781628955132 (ePub)
Subjects: LCSH: Rhetoric. | Rhetoric—Philosophy.
Classification: LCC PN179 .I64 2023 | DDC 808.001—dc23/eng/20230308
LC record available at https://lccn.loc.gov/2023001798

Cover design by Anastasia Wraight
Cover art is Mountain background, Stock Vector ID 2065813283, Net Vector, Shutterstock
Back cover: NOAA Rapid Refresh Smoke Visibility Imaging for North America on July 22, 2021. (U.S. Department of Commerce, National Oceanic & Atmospheric Administration: Global Systems Laboratory.) As part of John Ackerman's methodology and thesis, weather mapping in the atmospheric sciences coincides with racial 'weathering' if a reader's interpretation allows for it. One key point of this book is to propose that weather, rhetoric, and (for much of the volume), antiracist thinking and imaging are coeval because they are in daily living. John settled on the NOAA Hi-def Rapid Refresh imaging (featured in New York Times reporting) because it can be indexed to a specific period of social unrest. The summer of 2021 was a violent period of police violence and BLM and other protests, the aftermath of the January 6 insurrections, and yet another season of global weather events in the central Rockies (fire, floods), where the author lives and writes.

Visit Michigan State University Press at *www.msupress.org*

Contents

Acknowledgments

There are many things to acknowledge when writing a book with many authors. It's important to acknowledge, for instance, that each of us did this work on lands that are not our own. Each of us has had the *privilege* to do this work because of our professional appointments at universities where we were hired and others were not. That means we would like to acknowledge others living in material conditions that are less livable than ours and for whom livability has nothing to do with research or writing. We would like to acknowledge that sometimes it feels like the world is broken. And: not despite this, but because of this, we need to acknowledge that there are still affirmations worth making, delight worth taking, amid the air of all the brokenness. We would like to acknowledge Catherine Cocks and the entire staff at Michigan State University Press for helping to bring this multigraph to life. We would like to acknowledge our reviewers and blurb writers for their generous labor. We would like to acknowledge the essential workers. We would like to acknowledge our families, our kin, and our support systems. Finally, we would like to acknowledge and affirm one another. In different ways, we have found shelter in one another throughout this project, and that which is precious should be acknowledged.

A Note about This Book

This book was born from a book club—though our first intention was to read them, not to write one. That changed after three years of Zooming every two to three months when we decided to form a conference panel in order to share some of the thinking-feeling we'd been doing together over that time. The provocation that spurred us wasn't any one of the many crises unfurling around us: fires in the streets, fires in the forests, racism everywhere, the rich getting richer, extremism, COVID. Rather, the open-ended prospect that thinking in terms of "climates" might be a salient way to make sense of rhetoric's being-doing in the world struck us as particularly promising at a time when the persuasive actions of individuals seemed insufficient to explain the sense that emergent *conditions for change* were blooming and ambient all around us. We weren't then, and still aren't now, quite on the same page about the virtues or possibilities of "climate" as a concept. In a key way, though, this undecidability was and remains its conceptual allure. It makes this book fundamentally exploratory and speculative; it represents both individual and collective reflections that cross, align, and depart from one another. In practice, we wrote each of these chapters individually, but then revised collectively, adding on, subtracting, exploding. As transitional, reorienting seams between each chapter, we've added short microclimates written collaboratively by the two people who wrote the chapters surrounding them. These explore how ideas in the chapters before and after come together or clash, creating their own unique conversations. Throughout, we've also responded to each other's work in conversational notes inspired (because we're often inspiring one another) by something that one of us wrote. These notes can be distinguished from more citational notes by our initials being in superscript. By way of ending, we've pulled back the curtain, offering a lightly edited transcription from a meeting of our reading group, in which we discuss, well, a book called *Rhetorical Climatology*.

Our Reading List

Waves of Knowing, Karin Amimoto Ingersoll

The Brain's Body, Victoria Pitts-Taylor

The Spell of the Sensuous, David Abram

What a Body Can Do, Ben Spatz

Animacies, Mel Chen

Friction, Anna Tsing

The Hundreds, Lauren Berlant and Kathleen Stewart

In the Wake, Christina Sharpe

The Black Shoals, Tiffany Lethabo King

Demonic Grounds, Katherine McKittrick

Wayward Lives, Beautiful Experiments, Saidiya Hartman

Ornamentalism, Anne Anlin Cheng

Queer Inhumanisms, Dana Luciano and Mel Chen (editors, Special Issue of *GLQ: A Journal of Lesbian and Gay Studies*)

Native American DNA, Kim TallBear

A Billion Black Anthropocenes or None, Kathryn Yusoff

Pollution is Colonialism, Max Liboiron

The Economization of Life, Michelle Murphy

Introduction **What We Talk about When We Talk about Rhetoric**

Hot air, of course. Spin. Misdirection. "Strategery." The seeking of advantage through fluff or flourish. Let us not forget: such things are what most people talk about when they talk about rhetoric. But then, most people don't talk about rhetoric at all. They use the word (*if* they use it) as a descriptive label, like "pink" or "crunchy," to stand in for all the verbal artifices that people deploy to get what they want. This means that in everyday contexts, to identify something as rhetoric is already to identify it as biased, or at least partially suspicious on account of its partiality. Such suspicion is nothing new. A certain distrust has attended the study of rhetoric since its outset, when Plato disparaged the art as mere cookery. The basis of such originary caution is often cast as concern that skillful rhetors, by virtue of their persuasive powers, might deceive or mislead others in dangerous ways. But that basis is itself premised on still another assumption, one so fundamental as to go virtually unexamined in the first 2,500 years of rhetorical studies: the baseline presumption that rhetoric is a human art, and therefore originates in the volition of individual actors, typically through speech or other symbolic means of persuasion.

The last several years have seen scholars in rhetoric beginning to operate from different presumptions. One reason is the increase of scholars inspired by the wisdoms of non-Western and Indigenous traditions, which have long been advocating for more relational, interconnected, and more-than-human understandings of language and meaning-making. Rather than understanding rhetoric to originate and reside in humans, those who have pursued such approaches often take rhetoric to be more expansive, even inherent to existence itself. Rhetoric, that is, pervades being. *Rhetoric is climatic.* That hardly means "it" isn't many other things as well. Most of those who study rhetoric today, whether by profession or predilection, would acknowledge that the term refers to a practical art and theoretical lens that has been an integral part of public life and education for millennia. While we have no designs to disagree, there are many reasons to shed some of the received histories and anthropocentric assumptions that continue tethering rhetoric to individual human actors and symbolic action in bounded situations.

To begin with, failing to do so leads to hideous exclusions: overlooking non-Western traditions and epistemologies; entrenching the privilege of "good men speaking well" at the neglect of women and many others who don't fit the cis-male-het-White standard; let alone eliding more philosophical concerns about intentionality, agency, the nature of semiosis, and so forth. Maybe above all, to continue accepting the field's unstated but predominant first premise—that rhetoric is an art of autonomous human subjects, acting symbolically in bounded situations to exert some effect on fixed audiences—will mean that the dynamic and worlding force of rhetoric will always be taken as either a peril or a promise. It might, as it has for many of those keen to recognize rhetoric's power, be elevated to a means of constructing realities, for better or worse. But even that effectively sidesteps the most important questions: *Better or worse for whom? Better or worse for what?*

The false dichotomy of rhetoric as danger or deliverer belies the material complexity of its both/and tendencies. There is no need to marshal the many, many definitions of rhetoric that have been offered in the long history of the term (often in neat, italicized words), to show its multiplicity and multiplication across milieux. There's a simple reason that "rhetoric" has always been so variously treated and defined: it's inherently a manyfolded thing. The beauty behind the "rhetoric of rhetoric,"[1] as Wayne Booth liked to call it, draws us to the very word's dynamism, its way of resisting definition or containment. This multiplicity is not just a matter of arcane academic disputation or quibbling differences in opinion. As Nathan Stormer has argued, rhetoric is ontologically "polythetic." In other words, what gets called "rhetoric" in actuality describes a whole class of coexistent things that have many, but not all, properties in common. Rhetoric, that is, "is not one body with many faces, but homologous affordances materializing many bodies."[2] The materializing of bodies, the worlding of culture, the politics of hierarchy—all are part of rhetoric's purview and, to put a rather loose-fitting bow on it, the subject of this book.

The "in Which" of Rhetoric

The important thing about rhetoric, after all, is not what it *is* but what it *does*. And at least when it comes to the study of rhetoric in American educational institutions (including the English, writing, and communication sides of campuses), it is hard not to notice that studying rhetoric these days is increasingly less about bespectacled analysis of texts and talk from a distance, and more about the shoes-and-socks work of engaging with actual people, communities, and environments "out there" in the mash and muck of life. One manifestation of this trend is the impulse toward more activist-oriented work, which sees the study of rhetoric not only as a generative lens for critical analyses of public affairs but as coterminous with the work of actually redressing unjust or unsatisfactory social circumstances.[3] Exceeding a long-standing belief that the birthright of those studying and teaching rhetoric is the development of competent citizens, activist approaches to rhetoric move outside the classroom, investing deeply in communities and people to help serve the achievement of their needs.[4]

A parallel expression of this trend can be seen in the growth of "participatory rhetoric," and its various field methods, as a way to attune to the dynamism and complexity intrinsic to those communities or other scenes of investment that are best understood only when experienced through embodied, social, and diachronic involvement among them.[5] The turn to field methods has called attention to the shortcomings of imagining that the work of studying rhetoric could ever be wholly autonomous from the somatic and affective engagements that animate that study to begin with. To some degree, then, all rhetorical methods are already field methods because it is impossible to observe "rhetoric" without also participating in complicity with it.[6] From this standpoint, the ethnographic model of "participant-observation" is the ultimate redundancy, each being impossible without the other.[7] No wonder, then, that so much of the writing about rhetorical field methods has involved metacommentaries about how to undertake such methods. Like the activist-oriented impulses charging recent scholarship, the value of actually practicing these field methods cannot be reducible to their published outcome—including in a book like this—because it's primarily through the experience of being with and attuning to others that material *change* is prefigured and sometimes actuated in practice. What does rhetoric *do*? According to this thinking, it moves change.

In this book, however, we ask what it would mean to think about rhetoric not only as a *mover* of change but also as one of the *conditions* in which change occurs. Certainly, if rhetoric is a potential mover of change, to suggest a de-emphasis of that force is to risk seeming to endorse an unjust status quo. During the lifetime of everyone currently alive on the planet, human beings have never been more connected through technologies, supply chains, and means of travel, nor more separated by inequities: racial, religious, sexual, economic, and more. Undeniably, the globally dispersed failures of human rights, environmental justice, and equity, particularly for non-White people and those in poverty, are enormous and urgent. There are compelling arguments to be made, in other words, that all of us engaged in rhetorical research—really, everyone alive—should drop what we're doing and join the cause, *some* cause, to level power and bring about better social circumstances for those who are victims of racist or otherwise prejudicial policies.

But to treat rhetoric principally as a mover of change is also tacitly to constrain it as an (anthropocentric) instrument of dominion. This means that to study rhetoric as an activist means of achieving social or political change might itself risk effecting such change on the very back of an instrumentalist, individualistic, and patriarchal logic that may well be what needs escaping to begin with. As Audre Lorde is well known for cautioning, "the master's tools will never dismantle the master's house."[8] Maybe it's best, then, not to think of rhetoric as a tool but as something closer to the house itself: the material-discursive conditions in which lives happen and which contribute to determining the possibility of change in the capacious context of shared existence under the roof of a single planet's atmosphere. It might, that is, be illuminating to think of rhetoric as climatic.

From Ecologies to Climates

Before we can get to the many ways that it makes sense to conceive of rhetoric as climatic—and the study of rhetoric, then, as a kind of climatology—let's go back to that metaphor of a shared planetary house. While we all share a planetary home, clearly we don't all share it in the same ways. To take just one example, the study of (and struggle for) "environmental justice" shows that the deleterious effects of resource extraction, pollution, and other environmental harms disproportionately impact people of color in economically poor communities. That part of town near the toxic dump, or near the cancer-causing power plant, the area with unsafe drinking water? They're likely *not* to be in wealthy neighborhoods. But globally, too, the "global south" fares far worse than the "global north" when it comes not only to opportunity and prosperity but also to innumerable ecological vulnerabilities propagated primarily by those northern superpowers that, taking planetary degradation as their prerogative, act with disregard for its downstream consequences. In other words, while all humans (and far more than that) unavoidably cohabit the same planet, not all of our respective positions on it—from our individual neighborhoods and homes to our geographical location on Earth at large—can take for granted the same freedoms and privileges. In this context—one that spans the local and the global, the individual and the collective—how can we describe the sharedness of our common home?

It turns out that there's already a trenchant word to describe the household of organisms on Earth. A German zoologist named Ernst Haeckel coined it, in 1866, when he described "oecologie" (ecology) as "[der] Haushalt der thierischen Organismen"—the household of animals.[9] It can be helpful to contemplate what it implies that *oikos*, the Greek basis for the term, is also a word that figures at least indirectly into how rhetoric has been understood since the Greeks used the word so long ago. As Hannah Arendt has shown, if *oikos* was the private realm of the household, its counterpoint was the *polis*, the public realm of political activity.[10] As anyone who studies rhetoric would know, it's the *polis* with which rhetoric has seminally been concerned, insofar as public-facing human discourse has long been held as the paragon of political responsibility in the belief that eloquent speech can guide people through the challenges of living together with others despite their different prerogatives and opinions. In rhetoric's very association with democracy and public affairs, it has always been marked by its separation from the *oikos*.

Arendt's argument about this separation aside, the etymological point is that ecology has from the start been regarded as "the science of the household of organisms, i.e., of their relation to their biotic and abiotic surroundings."[11] But if the Greek distinction between the *oikos* and the *polis* gives us the very basis for theories of public life, hence of rhetoric itself (the action of speech being, as Arendt puts it, "what makes man a political being"),[12] then this newer concept of ecology rehabilitates the idea of *oikos* with different connotations.

Notice here that ecology's seminal concern with the "household" might connote private spaces—the den of the bear, the nest of the bird, the rabbit's burrow, let alone human households—and also the natural world itself, under whose celestial roof all things coexist. The rehabilitation of the Greek concept of *oikos* at the very advent of ecological thought, in other words, enlists it for a decidedly different purpose than to reinforce the ancient divide between the private sphere—home to the free but essentially hidden interactions of the domestic realm—and the public sphere, which accommodates the open interactions of free citizens in the political.

If there were ramifications for the study of rhetoric after Haeckel's advent of "ecology" as a named concept, they were not immediately felt. He was not, after all, a rhetorician so much as a Darwinist, and in the 1860s (and until much more recently than that), rhetoric remained largely associated with a public/private divide and the role of speech in public life. Our aim in this book is not to write rhetorical history, so we aren't going to offer a story of how ideas about rhetoric have changed since the naming of ecology as such. What we will say, however, is that although the entrance of ecological thought into the study of rhetoric was virtually nonexistent for over a century, that is no longer the case. Delayed or not, the consequences of introducing an ecological perspective into rhetorical studies are quite radical—and have inspired our turn to "climates" as a specific approach to thinking ecologically about rhetoric.

What is it, then, that we talk about when we talk about rhetoric? Many things, of course, depending on whom you ask. But a core tenet of the climatic approach that we endeavor here is the tendency to resist partitioning rhetoric in a hierarchical taxonomy of topoi and tactics, even as its very force-effects often manifest in the hierarchies of being that rhetoric itself builds and maintains. One thing that might follow from such an ontology is an ethical charge for rhetorical scholarship to expose these hierarchies by treating the "climates" *in which* rhetoric exerts a force on the world *as themselves a rhetorical force* on the world. Such an endeavor offers a fundamentally ecological way of thinking about rhetoric because of its baseline presumption that all things exist in relation to others: intrinsic, unavoidable, even radical relationality is ecological thought's sine qua non.[13] But because the fundamental relationality of ecologies does not imply an *equality* of these relations (romanticized popular understandings of ecology to the contrary), we turn to "climates" as at least a semi-stable way to bind the unspooling, endless relationality of "ecology" in general, into somewhat more coherent scenes and sites of rhetoric's ambient force. This framework allows us to make what Karen Barad calls the "agential cuts" that, as Sofie Sauzet summarizes, "enact that which is inside and outside of phenomena in a single movement."[14] Our wager is that emphasizing climates accordingly will show that rhetoric's force as a potential agent of change is always inseparable from its force as that set of ambient conditions in which change does or doesn't occur.

It's not incidental that this move comes at a moment that feels like a disciplinary reckoning, a coming to terms with what the work of rhetorical scholarship involves (or should involve) in the face of so many mounting real-world failures in broader human relations, both with our more-than-human kin and with the planet itself. By building from, but going beyond, poststructural investments in critiquing Western humanist ideals of free, rational, and autonomous human subjects, ecologically inflected thinking about rhetoric rightfully emancipates rhetoric from hoary notions of good men speaking well in situated contexts. In turn, those inclined to such thinking have a much broader and more complex range of factors to consider when thinking about the rhetorical influences that potentiate, precipitate, and produce legible change in the world. Given this range, we've found that thinking in terms of rhetorical "climates" offers a particularly apposite way to make sense of those ambient but ecological milieux that set the conditions and consequences for all that transpire within them.

The "We" of Rhetoric

Prior to what we're trying to do here with "climates," perhaps the most prominent way ecological thought has influenced the study of rhetoric involves its impact on the ongoing reconceptualization of so-called rhetorical "situations." Specifically, in an important article from 2005, Jenny Edbauer describes rhetorical ecologies as "co-ordinating processes, moving across the same social field and within shared structures of feeling."[15] But other key concepts have been equally challenged and charged by the introduction of less bounded ecological thought. In many regards, the turns to space and place in rhetorical scholarship, and more recently, to temporality, attest to the growing recognition that broader material-discursive scenographies of interest are necessary if the aim is to understand the ways rhetoric operates as a force in the world. That force is ecological in its emergence, but climatic in its provisional stabilizations. Rhetorical forces are born from the chronotopic relations that make them possible and, accordingly, correspond with the many spatial and temporal forms that compose newness from within conditions that were once new but are no longer.

In one of the seminal and most focused works on the implications of thinking ecologically about rhetoric, for instance, Nathan Stormer and Bridie McGreavy suggest that such thinking calls for new ways of conceiving some major commonplaces of theory in the long tradition of rhetorical studies: "from agency to capacity, from violence to vulnerability, and from recalcitrance to resilience."[16] Sid Dobrin's work, across several books, has long emphasized the ecological nature of the rhetorical, particularly in written composition.[17] More recently, Joshua Trey Barnett has offered that ecological rhetoric is not, or at least should not be treated as, just a niche concern of those rhetorical scholars whose research focuses on "environmental matters." To the contrary, ecological rhetoric is a project of studying and cultivating "earthly coexistence," a multispecies endeavor that requires *ecocentric* modes of thinking and relating.[18]

Even if research undertaken overtly in the name of its ecological commitments remains, as yet, just one among many of rhetorical scholarship's generative lenses and objects, the last few decades across the field have certainly seen "the rhetorical" set loose into material-discursive ecologies of circulating, ambient rhetoricity. The clever sound-bite version might say that the field is becoming increasingly attentive to *pre*suasion in addition to persuasion. If you'd like more evidence of such attentiveness, there are many places to look, but good starting points include Thomas Rickert's work on choric ambience, Diane Davis's work on "preoriginary" obligations toward the other, Chris Ingraham's attention to "gestures of concern," Megan Eatman's work on rhetorical violence in *Ecologies of Harm*, even in more historical research, by Debra Hawhee, to uncover some ways that sensation has always been central to conceptions of rhetoric.[19] In the form of a question, such work might be said to ask, *What different conditions of possibility create different kinds of persuadabilty*? This is quite different than the old and tired project of asking after the available means of persuasion. The inquiry rather concerns what makes some means available and not others, and *that* concern involves interrogating and challenging the structures of power and hierarchies of being that value and privilege some existents over others.[20]

What we are trying to suggest is that the expansion of rhetorical analysis to its more climatic and ecological registers—which entails greater attention to the relationships between conditions for influence and means of influence—is also one that entails raised attention to unequal and inequitable judgments about what matters, and who matters, relative to different ambient circumstances. And the ambience here is crucial, because ambience, like ecology, is a kind of figurative household, independent of any single person yet influencing all that it envelops, human or not. Of course, we should need no reminding that, in a time of planetary collapse, the figurative household that we all share is this planet, Earth, beset by a heating climate that not one of us is responsible for, yet not one of us alone can fix. The grand cliché of ecological thought is that "everything is connected to everything else." That's well and good, but the platitude elides the messiness of the ways these connections are in practice disproportionate, unevenly distributed, and capable of being thrown out of whack or severed altogether.

Perhaps, then, there is some benefit to thinking about rhetoric as a dispensation to determine the *we*. Who counts as *we*, after all, is often what "we" talk about when we talk about rhetoric: Who's in and who's out, with whom to share and from whom to withhold. And not just *who*, but *what*. What matters? What's true? What to do? At least since Kenneth Burke elbowed persuasion off-center to spotlight identification as rhetoric's first business, most of us who bother to talk about rhetoric have been thinking, in one way or another, about how assent to any "we" occurs. This would not be important if "we" were not such a dangerous word, a presumptuous one, worse perhaps than the cruelest of exclusions because it condescends to circumscribe others within the ambit of personal experience that may bear no likeness to one's own. And yet, aren't these times—this small and finite historical span in which the living share time on the planet—also times marked by phenomena so vast, so encompassing, so *global*, as to implicate everyone? How without loss to account for the commonness of contemporary experience in face of its sheer plurality? And how, inversely, to acknowledge and celebrate the vast diversity of difference without denying the real commonality that people do share?

Working through *Rhetorical Climatology*

The work of rhetorical climatology offers one potential response. Yet, our aim in this book is not to offer a theory of rhetorical climatology or a primer to help operationalize it as a stable and cohesive methodology. And still less should these pages be read as a manifesto propounding some core axioms that we share. We are not looking to recruit researchers to a program or sell readers on a particular way of doing rhetorical scholarship that we happen to find generative. To do so would instrumentalize a hierarchy of importance—*this* way is the best way, *this* is the work that's most necessary—and imply a privileged center that would undermine the whole project. What we call rhetorical climatology simply seeks to account for what's missing when the ecological dynamism of rhetoric is not taken as a starting premise. In that sense, this book offers a new way of thinking-feeling-doing research in rhetoric by seeking to enact the collaborative and mutualistic spirit of a "we" that is always still in formation and at risk both of excluding others from its sharedness and including those who might want no part.

Consider, for instance, how the *we* that the six of us have built over several years as a reading group differs from the *we* that we now own as "A Reading Group," forever more positing some cohesion on the cover of this book. While each of us have shared a broad alliance around common interests from the start of our first meetings—interests that we never discussed outright, though they could, if pressed, probably be described as a vague combination of some Rhetoric + Ecology + Equity + Intervention formulation—it would be inaccurate to say that what's held us together has been a specific topical focus or, say, set of methodological commitments. Curiosity can be a commons sufficient to itself. Couple that with the pleasure of engaging with those who share it, and our meetings have continued, even generated this book, because we've found them generative and fulfilling. Full stop.

Readers should not, however, perceive this to mean that *Rhetorical Climatology* is always a unified expression of aligned beliefs about, or approaches to, the problems that we're setting out to explore. Rather, we ask our readers to see each of us individually, and all of us together, as *working through* problems.[21] And we mean this in two registers. In the most immediate sense, everyone who works on anything, academic or not, is also "working through" the many challenging conditions of their life: paying the bills, getting food on the table, navigating the internal battles, the social dramas, the jeopardies and vulnerabilities, the too much of it all, against which what's conventionally taken to be "work" can sometimes feel separate, "extra." No scholarship anywhere exists without its authors having persevered to transcend—that is, to *work through*—those personal and interpersonal circumstances that are too immediate and pressing to be "extra" to anything. During our time as a group, more than one of us has lost dear loved ones; faced health and other emergencies; felt overwhelmed by all the care-providing, the work obligations, the personal demands of COVID, among the many other traumas unfolding around us. At different times during our collaboration, some of us were able to continue working only because of the grace, care, energy, and good will of the group. We worked through it together. All of this might seem beside the point, but the significance and meaning of this book is wrapped up in its making. Beyond any specific content, argument, or scholarly conversation we engage, in other words, the making of this book itself matters as a call for, and enactment of, more ethical, relational, intimate, collaborative, sustainable, and humane ways of doing scholarship—separate yet tethered, as people experiencing a shared world, but doing so within their own individual lives.

In its other register, however, "working through" a set of problems means modulating the expectations that they could ever be resolved or eradicated in full. We do so here without the scientific ideal of "solving" anything, but rather with the inclination that working through problems, toward an end that may never be reached, is itself a responsible orientation to scholarship because it refuses to posit a conclusive telos, resolution, or "outside" to any enduring trouble that would neglect our own complicity in its perpetuation. In this way, *Rhetorical Climatology* sheds its argument as it goes along, like an elk shedding its rack, or a sudden movement of clouds shedding a light that may as soon go away. Readers expecting an argument to be built in linear aggregate, one page or one *claim/datum/warrant* to the next, miss that, when such an argument is finished being built, the implication can be that there's no more work to be done. "That about does it," says the finished argument. Here, by contrast, the ongoing work is what we're asking readers to contemplate, feel, or see. And such work entails a constellation of wonderings, arguments, stories, emotions, actions, and shoaling insights that emerge only here or there, in threads and refrains, in intimacies, ties to scholarship and the experiences of everyday life within our social climates. This approach may not oblige the trained academic reader's demands for ratiocinative arguments with nifty synopses rendered useful through their instrumentalization, the way heat renders fat down to bare bone, leaving something cooked but calcified, something denuded of all its energy—but easy to identify: "Those are the bare bones of the argument." This book is energized by the problems it can't just render away.

But what exactly are these problems? A quick survey of the ambient context for the first three years of our meetings offers a decent sketch of their contours: Black Lives Matter, #MeToo, Trump, Brexit, conspiracies, refugee crises, mass shootings, rampant racism, insurrections, COVID, climate change. *What in the world was happening*? From around 2018, when we first began meeting, at least in America something more encompassing and enduring seemed to be settling in. It was hard not to feel that all the ambient atrocities signaled an entrance into a new kind of "climate," social and meteorological alike, but in either case, one that nobody asked for and nobody could escape. What was its nature? And how could the study of rhetoric—even in its broadest, most rangy multiplicity—help us to understand this climate more clearly? These were problems we found worthy of exploration. Worthy of a whole book. And even if we've never fully agreed upon their solutions, working through them has become an anodyne for their frustrations, a way of opening a futurity, albeit one unspoiled by undue optimism.[22]

More Than Me, Less Than We

As we use the term here, "climates" can be understood as the long-term pressures of global give-and-take that bracket some possibilities while making others more materially and rhetorically accessible. Although that give-and-take is often most visible in relation to the symptom, or the puncture, its ecological structure enticed us. This collaboration is accordingly born from an honest attempt to bring those undercurrents into focus, and to describe, theorize, and critique the climatic conditions that support the emergence of certain symptoms again and again. Especially for rhetoric, a field that tends to highlight the situation, the specific rhetor, the ecological joint, we found that the metaphorical *and* material realities of "climates" offer a crucial way to understand the oft-illogical interconnectedness of events, beyond individual human action but still shy of full collectivity.

This book, then, is neither a monograph nor a gathering of edited essays. By design, it falls somewhere between the two: on one extreme, the romantic vision of a sustained, self-consistent, and unified endeavor by a single mind; and, on the other, the familiar patchwork of an "edited collection," the kind framed earnestly enough by a cohesive theme, while in practice the contributors work without much interaction at all. The in-betweenness of this book means we have worked alone and together in many ways and over long periods of time. But we have always interacted regularly, both through scheduled group meetings and in the aleatory ways whereby inspiration sparks spontaneous offshoots of conversation between friends. *Rhetorical Climatology*, in other words, is a multigraph. It is multiply written. This means it can be understood in the same, if minor, tradition of other hybrid collaborations-and-individual works, such as Miyarrka Media's *Phone and Spear*, or The Multigraph Collective's *Interacting with Print*, among others. In our variation, the narrative "voice" sometimes shifts from "we" to "I" and back again and sometimes avoids the distinction at all. Being part of a group, being invested in a multigraph, means thinking multiply across our differences: "More than me, less than we." We are multiple. Rhetoric is multiple.[23] That's what we're trying to do and model.

For instance, we've encountered the multiplicity of our difference in the uneven demands, pressures, and privileges we experience in our academic lives and how these shape abilities to make time to read, especially during already busy semesters and when we were also adapting to teaching and living online. Of course, we've also experienced differences in our respective (dis)abilities; sensibilities; dispositions; and racial, ethnic, and gender identities as well as in the relative comfort, ease, and accessibility of using online technology to meet virtually. We've encountered difference in terms of what a multigraph book project means at the respective stages of our careers. All of these differences and many more besides shape our collaborative method, which is more a heterogeneous and emergent set of practices and collective activities than a formal, organized process that another group could follow. This process-as-method has allowed us to learn from each other, tap into sources of support that we each needed at different moments along the way, and be in dialogue as a means of identifying what "the field" and "the world" might be asking of us as scholars, but also as people with lives beyond our professional identities.

As a result, what cohesion our book attains across its pages comes from much more than the commonality of many hours same-paging our conversations together. In practice, our compositional process involved each of us initially writing individual chapters on our own (denoted here by our first name in each chapter's heading), and only subsequently collaborating together. But even this "on our own" was informed by years of reading, thinking, and feeling together. So, while there are numerous concrete examples of collaborative writing we can point to—such as pairing up to cowrite the transitional "microclimates" between chapters, crowdsourcing read-throughs and revisions of the whole manuscript, interjecting throughout the book with conversational notes of our bantering voices, even offering a transcribed dialogue as our "inconclusion"—it is also true that many of our "collaborations" are harder to pinpoint. For example, while inquiries that emerged through conversation certainly get picked up in this book, they've inevitably done so in ways that not all of us are all aware of, in part because they could have blossomed from an exchange that only involved two or three of us. These and other unknowable influences have shaped the collective meaning made here, as well as our shared capacity to make this together to begin with. This "more than me, less than we" nature of our methodological process has been one of its great challenges and pleasures.

One inspiration for such a project comes in the final pages of Anna Tsing's magisterial book, *The Mushroom at the End of the World*. As an "anti-ending" that refuses to proffer a tidy that's-all-there-is-to-it conclusion, she suggests that the best intellectual work requires "playgroups and collaborative clusters: not congeries of individuals calculating costs and benefits, but rather scholarship that emerges through its collaborations."[24] If methods are traditionally understood in the "tool-belt" model, whereby methods are tools used to produce knowledge, then it's necessary to know in advance what kind of knowledge one wishes to produce so that the appropriate tool can be selected to produce it. If you're trying to catch a butterfly, you wouldn't grab a hammer. But if scholarship emerges through collaboration, by working through things with unknown ends, then it has its own methodological rigor, adapting as it goes along. Collaboration, not individuation, is just what's needed to move beyond the tidy, the scalable, the measurable, and the myth of heroic human individualism as a means to solve collective problems that in actuality require collective solutions. *Rhetorical Climatology* tries to do so within the interdisciplinary study of rhetoric.

Conventionally, of course, our methods could be described more generically as "mixed": a blend of historical research (chapter 1); rhetorical theory-criticism (chapter 2); digital ethnography (chapter 3); *in situ* field work and visual rhetorical analysis (chapter 4); place-based, antiracist institutional transformation and storytelling (chapter 5); sustained but unstructured conversations with local police departments and stakeholders (chapter 6), among others. But to emphasize these tools as essential to the production of this book—however familiar they may be, however intentionally each of us may utilize them—misses the crucial affective labor of thinking-feeling together about the challenges of living meaningfully in a time when the planet is collapsing, democracy is in peril, and hideous exceptionalisms keep holding people back from their flourishing. And these are the major topics at hand. We're just trying to work through them.

Each of the six chapters ahead accordingly offers its own variation of thinking-feeling a rhetorical climate: deterministic, racist, ableist, environmental, institutional, and violent. These hardly constitute a comprehensive typology of rhetorical climates (as if such things could ever be enumerated and organized into a finite classification scheme). Instead, we treat this work as offering different ways of looking at the climatic nature of rhetoric when "it" is not held to have an isolable and discrete origin, locus, or motive. Though the largest shared context for these ways of looking is undoubtedly the looming specter of mass extinction posed by climate change, none of this book's chapters are "about" climate change as such. Rather, as Candice's chapter emphasizes most explicitly, they are about change-climates and climate-changing labors: the worldbuilding ambience of rhetoricity and the possibility for things to be otherwise than they are. By staying adjacent to the haunting grief and fear of total planetary collapse, we seek to show that in the alleys of the everyday, in the contours of personal experience, in the crescendos and disappointments of social life, in communities, on screens, among kin, or with strangers, mutualism is all around—if only we bother to notice.

1 Digression on Air (Chris)

How can a book begin with a digression?[CI] What could its opening lines possibly be digressing from? And a digression on air, no less, a medium so ethereal that it's visible only through the impurities within it. But we begin in the air all the same. Not for the vantage that comes from being carried aloft, surveying from on high, but for what's harder to see: the air itself. For in so many ways, this book is about the air, about the weather, about the "climates" that we can feel and sense, but not always see in any empirically legible form. Air is that in which all terrestrial existents, living and not, go about their daily offices. Clearly, the invisibility of air does not make it less pervasive or less integral to planetary life. Indeed, the simultaneous invisibility and pervasiveness of air contributes to its absolute taken for grantedness. And the taken-for-granted is the always-here that isn't. To attempt identifying the taken-for-granted, to see things *as* taken for granted, is therefore always a digression: a temporary departure from something so substantive that, most of the time, we don't pay it much mind at all.

In these days of extraordinary precarity, it's harder to take anything for granted, the air being no exception. As fossil capital produces particulates and pollution that hamper human respiration, the air becomes more visible, harder to forget. The pandemic, too, has made the air more suspicious, carrier as it is

CI Hello. Chris here. Glad you could join us. This is where the book's hidden transcript lives. Whenever you see a notation with CI in superscript, that's me, chiming in. We all have our own superscripts with our own initials so you know who's talking: JA = John Ackerman, JL = Jennifer Lin LeMesurier, BM = Bridie McGreavy, CR = Candice Rai, and NS = Nathan Stormer. And we do a lot of talking here. Agreeing, amplifying, pushing back, counterpointing. These are different from the numerical notations you'll also find throughout, devoted to citations and the occasional authorial comment. So, if you see our initials in tiny floating letters, that means there's a conversation that you don't want to miss. Sometimes we even have jokes. Like, *How do you keep a reader in suspense?*

of the invisible airborne particles and respiratory droplets that have demanded our masking-up. But the air can also be a carrier of moods more impalpable, from Shakespearean airy nothings to whatever Phil Collins feels coming in the air tonight. If the palpable intensity of the American summer following George Floyd's murder in 2020 is any indication, having air taken away can leave revolution in the air instead: "I can't breathe."[1] Of course, while intuitions of something as notional as an emergent social mood can well be real, they are often baldly personal and then projected outward as if they were more universal. It's when the reverse comes to pass that our inexorable entanglement in a meshwork of others comes to the fore, because all too material perceptions of something doing "in the air" of the social field can also inspire new possibilities and urgencies—from outside-in.

The things we take for granted tell us a lot about the things we value and depend upon without thinking of them as values or dependencies. Yet, to take anything for granted is a privilege, and, like so many privileges, not typically acknowledged at all. One thing many people take for granted is the "we" itself: the vague sense of community and commonality that allows people of any kind to imagine themselves aligned with others by virtue of important similarities. But not all similarities are the same—being a mother, say, or being White, living in China, or having a distaste for olives—and to share some similarities with others does not mean sharing full similitude. The common is never *completely* common.[JA] To digress on the air, then, far from just to question the privilege of having clean air, or of not being choked on a sidewalk, rather involves asking after the ways that personal taken for grantedness can inculcate a sense of collective commonality that is sometimes imaginary, sometimes actual. To digress on air is to begin adrift, and to grow downward from there.

According to a lovely anecdote, someone once asked Abraham Lincoln how tall he was. Lincoln responded (one may imagine with a mischievous grin forming on the corner of his lips), that he was "tall enough to reach the ground." He hadn't, of course, read the now classic children's book from a century later, *The Phantom Tollbooth*, in which some characters are born in the air with their

JA I might go further. The common is never common, nor is the public actually public. Enlightenment is much more than enforced darkness, at least for the global south, per anticolonialist thinkers like Mignolo. See Walter Mignolo, "Delinking," *Cultural Studies* 21, nos. 2–3 (2007): 449–514.

head at the exact height it's going to be when they're an adult, and then grow down toward the ground.[2] But Lincoln must have shared a similar sensibility, that becoming grounded was an achievement made possible only by being ungrounded to begin with. To be born in the air and grow earthward, to find one's footing, is not to abandon the air, as if that were possible, but to dwell in a more ambient and ethereal kind of grounding. Growth and change are better understood in this sense not only as directed *toward* something but also as *coming from* something.

In principle, the idea of a commons is the notion of "coming from" something shared. Sometimes it's a place, sometimes a set of beliefs, sometimes it's a condition of unlivability or hospitality or loss. Or many other things. But always it's the air. Like Plato's *chora*, the air is a way to conceptualize both that in which and out of which all activity occurs.[3] And what occurs in the air above all else is the weather. What better way to illustrate a mooding commons than with the weather that envelops all planetary being? Think of a rainy day in your town: overcast, wispy clouds slung low, necks hunching in horizontals over saturated sidewalks, a grayscape dappled with drops in puddles. It is less the case that such weather precipitates a common mood—moods, after all, being capable of differing from person to person—than that the sharedness of weather establishes some conditions of possibility for moodiness as such. And there is always weather, just as there is always moodiness. There is no non-weather, no affective neutrality.[4]

Still, the weather can only get us so far toward the ground, because the weather itself is a condition of atmospheric expression. Weather doesn't create atmosphere; it comes from atmosphere. But atmosphere is no final origin either, as atmosphere is an expression of climate, whence it comes. For the scientists working in America's National Oceanic and Atmospheric Administration (NOAA), the difference between weather and climate is a measure of time: "Weather is what conditions of the atmosphere are over a short period of time, and climate is how the atmosphere 'behaves' over relatively long periods of time."[5] This explains some of the reason that climatic temperatures can trend toward heating steadily over longer periods, while still undergoing smaller, intermittent periods of cooling along the way up. Unlike the weather, though, which manifests in numerous short term "weather events" around the planet, each of which can change by the hour and from place to place, and unlike the atmosphere, which is really just a semi-stable, layered series of differently composed gases held by gravity near the planet's surface, climates are harder to describe—despite being more important.[NS]

How we think and talk about climates matters. The trouble is that "climate" has become a word vested with the burden of doing too much work. In one moment, it refers to the subject of a planetary crisis like none ever known in human history; in another, it refers to a kind of public mood or atmospherics of sociality. Climates are both locatable and messy: they can be identified as distinct, but they also bleed one into another. Climates are both material and intangible: they're marked by measurable and empirical phenomena, but they can't be touched or contained. Climates are both predictable and erratic: they follow certain trends but also do what they want. Like global warming itself, each of us can contribute to climates, though it takes a multiplicity to make or unmake them.

For all these reasons and others, "climates" are thoroughly rhetorical constructions. Like all forms of meaning, what we talk about when we talk about climates will always be context-dependent, subject to revision, and contingent upon horizons of interpretability unique to any given audience and conjuncture. Yet, the ways people talk about or otherwise represent climates has a consequential influence on the array of real-world responses (and non-responses) that humans take toward different climates and any perceived needs to mitigate, redress, or otherwise deal with them. Readers here likely won't need reminding that climate-change deniers are still out there, refusing to accept the depiction, backed for decades by scientific consensus, of a planetary climate growing increasingly warm.[BM] The rhetorical constructedness of climates doesn't mean

---

NS I think a crucial aspect of the concept's challenges is that climates are ethereal yet palpable *to the skin*. Something we do not really engage with in this volume, I think, is the way that climate is understood dermally: heat, cold, wind, fire, water. Climate is felt through elements as they impact skin in the usual sense of its usage. One does not describe internal, material responses climatically; one uses poison or infection logics usually. A virus is not climatic but infectious. So, I wonder to what extent there is an analytic of surfaces that we are writing with and against?

BM This link to climate denialism and climate realities makes me think about how incommensurability threads through this book, and how it shapes my community-based work focused on climate and coastal change too. I work with several clammers who some would call climate deniers, though I don't use this term because they wouldn't refer to themselves that way. Many clammers accept climate science but many also don't. In some cases, this difference in perspective, or reality, is at least partly due to how, as fishermen, they have direct and repeated

climates aren't also real, after all, that they don't exist in actual ways, marked by specific identifiable phenomena and not others. Nor is everything a climate. Despite the capaciousness of the word's broad English usage, the many ways people talk about climates exerts a different register of influence than the many ways climates themselves operate to influence what they encompass. Whatever impalpable thing it is that the word "climate" refers to, in other words, is itself something with a rhetorical force.[JL]

All of this can be summarized in the form of a proposition: climates are rhetorical.[JA] And this is true in the two related senses intimated above. In one, they're rhetorical by virtue of the different ways that climates are discursively represented, including not just through language, but through science, images, instruments, folklore, artwork, and so on, each of which can construct climates in a particular light and with particular stakes. In the other sense, and at the same time, climates are rhetorical by virtue of the distinctive affective force that they can exert on both personal and social experiences of dwelling on this planet. These forces are not reducible to symbolic or signifying language but involve the "thing-power" of climates themselves as material-discursive assemblages constantly in formation, becoming variably influential on that which they encompass.[6]

If the word "climate" then refers to a host of different phenomena depending on the context of usage, and if these phenomena likewise correspond with different types of climates, then it may be useful to ask what all climates share in

---

experiences with how science is used as a powerful regulatory force, and sometimes based on evidence that is less than credible. And yet, in a context where "climate adaptation projects," or what we tend to refer to as "intertidal ecosystem restoration projects," are desperately needed now, we see time and again how clammers who distrust climate science are still digging in, sometimes quite literally, to lead climate adaptation and intertidal restoration projects. For me, attuning to rhetoric as climatological helps me remember incommensurabilities like these, and to be ready when the differences present opportunities for connection and disruption.

JL Thinking with disability here, some are also more sensitive to climatic changes than others, which adds to the complexity. Many disability activists are still self-isolating due to COVID risk. These bodies are micro-attuned to the rhetorical force of a climate that has recentered ableist metrics as "normal."

JA Wouldn't it be a lot easier to see how "climates are rhetorical" if asshat goon rhetoric were decolonized?

common despite their variability. Consider, for example, the generous difference between the phrase, "anthropogenic climate change" and "a climate of anti-Black racism." Each uses the word "climate" in a different way, though neither atypically. The first phrase refers, in general, to climates as something environmental, something with an air of the scientific, maybe something to do with the weather, carbon dioxide, or global ambient temperatures. The second phrase, meanwhile, refers to a different kind of climate altogether, one involving human interaction, emotion, attitudes, or beliefs and their sociocultural pervasiveness in a given time and place. It may be tempting to draw epistemic distinctions between the two, more or less along the old "two cultures" paradigm of the sciences and humanities: Climate A, related to prevailing weather conditions; Climate B, related to prevailing social condition. But the ability of words to denote or connote multiple meanings at once, sometimes even contradictory meanings, is not, after all, unique to the word "climate." The polysemy of language isn't the trouble worth staying with here.[7]

Where that trouble lies may seem to be with the proposition itself: if climates are rhetorical, so what? Aren't many other things as well? Speeches, photographs, monuments, menus, clothes, bodies. Anyone familiar with the academic disputations over "Big Rhetoric" in the 1990s would likely acknowledge the problems with identifying more and more things as rhetorical, until everything is rhetorical. As many who engaged these debates pointed out, if rhetoric is everywhere, it's nowhere.[8] There is no need to revisit the Big Rhetoric/Little Rhetoric conversations here, except to note that problems with the expansiveness of "climates" in some ways mirror the problems unearthed by the expansiveness of rhetoric.[JL] One need not imagine that climates are rhetorical to nevertheless imagine that climates are pervasive. To live on a planet is to be enveloped by a climate, no less than to live in language is to be enveloped by rhetoric. Of course, many rhetoricians would go much further than to limit rhetoric's ambit to signifying discourse. When we take rhetoric beyond human discourse—particularly in light of recent trends toward "thinking ecologically about rhetoric's ontology," as Nate Stormer and Bridie McGreavy have suggested in their essay by that name—it becomes apparent that the work

JL Given the Big/Little Rhetoric debates, it seems worth it to note that we are not advocating for "analyzing everything." Rather, thinking about rhetoric and climate together orients us toward questioning the concatenations of mood and affect that invest certain rhetorical artifacts with heightened interest.

of thinking about climates as rhetorical also invites thinking more carefully about rhetoric as climatic.[9]

This leads us to forward another proposition, indeed a core tenet of this book: rhetoric is climatic. Given the presumed need first to agree upon what counts as rhetoric before determining whether climates count as rhetorical, this might well have been our initial proposition. But that wouldn't do either, as the trouble then would have been effectively the same: to decide what counts as a climate before determining whether rhetoric counts as climatic. Mental somersaults aside, the hard-to-breach circularity of the problem is part of why in this book we are interested in exploring the co-articulation of these two linked concepts: rhetorical climates, climatic rhetoric. The project at hand, that is, involves what could be called "relationshipping" rhetoric and climates. In doing so, the object of study isn't one or the other, but the dynamic relation between them.

Take as an example this second proposition that rhetoric is climatic. The suggestion here is not *cladistic*, meaning it doesn't seek to identify a nested relationship between rhetoric and some "higher" phylum of climates, at least no more than the first proposition sought to mark the inverse order of relation. The chiasmus that at once treats climates as rhetorical and rhetoric as climatic attains its heuristic force through its fundamental refusal to impose hierarchy on either concept. Their relationship is *not* hierarchical, in either direction. The ways a given "climate" delimits and determines what can be and be done rhetorically within its reach does not make climates a controlling trope, certainly no more than rhetoric's animate ability to create climates marked by specific moods and possibilities (and not others) could be said to make rhetoric the superior actor. *Climates animate rhetoric* and *rhetoric animates climates*. To privilege one over the other is like supposing that the water or the bank alone makes a river when, in fact, there is no telling which is the greater force. The water and bank make rivers together.[10]

The circumstances whereby climates animate rhetoric and rhetoric animates climate are both material and discursive, atmospheric and social.[JA] In

---

JA Just playing here, but the fulcrum balance of water and bank belies the unequal decay of both. Liboiron shows the water now assimilates an arbitrary capacity for pollution; the bank now domesticates as a canal the unruly river. Just so fascinating, the imbalance of the decay, and I mean rhetoric to decay, because the weather will be, well, weather. See Max Liboiron, *Pollution Is Colonialism* (Durham, NC: Duke University Press, 2021).

other words, even keeping rhetorical climates to a simple dyad of two types—climates as weather conditions and climates as social conditions—begins to break down upon any closer inspection. Mike Hulme, who has probably written more about climates than anyone alive, makes the case that although climates have "scientific, political, economic, religious, ethical, and social dimensions," the idea of climate "needs to be understood, first and foremost, culturally."[11] To suggest here that it also needs to be understood rhetorically is not such a stretch, particularly given another "fundamental chiasmus" (in the words of Stephen Tyler and Ivo Strecker) "in which rhetoric structures culture and culture structures rhetoric."[12]

To put this differently, there is a long-standing cultural and rhetorical connection between the two kinds of climate at hand, meteorological and social. This means that any line demarcating these apparently different types of climate is blurrier than it may seem. It's not just that identifying two "kinds" of climate is an artificial convenience that reduces a multiplicity of climates into a tidy pair for purposes of simplification. It's that even the division that splits climates into those with generally meteorological and social characteristics is misleading because the two have always been entangled, since the very outset of studying climates themselves. To illustrate some of that entanglement, the remainder of this chapter wrestles with one of its most long-standing—and troubling—manifestations: arguments for climate determinism.

Inclining toward Division

Climate determinism (sometimes called environmental determinism) posits that the climate of a given physical environment influences the appearance, intelligence, disposition, and sociocultural trajectory of those living there. Instances of this belief find their way into all kinds of thought, Western and Eastern alike. Perhaps most famously from a Western tradition, Aristotle advances a position of climate determinism in his *Politics*, though without naming it as such. Building from Hippocrates's treatise, *Airs, Waters, Places* (400 BCE),[13] which drew sharp distinctions between European and Asian climates, including corresponding distinctions in the appearance and demeanor of their respective peoples. Aristotle also linked climate to human behavior—but added the explicit link between human behavior and human value.

People from cold climates like those in northern Europe, Aristotle declares, are "full of spirit but somewhat deficient in intelligence and craft knowledge. That is precisely why they remain comparatively free, but are apolitical and incapable of ruling their neighbors."[14] By contrast, he finds those people from warmer climates, such as Asia, "have souls endowed with intelligence and craft knowledge, but they lack spirit. That is precisely why they are ruled and enslaved."[15] Unsurprisingly, then, when it comes to those climates in-between the cold and hot extremes, Aristotle looks no further than his own backyard:

> The Greek race, however, occupies an intermediate position geographically, and so shares in both sets of characteristics. For it is both spirited and intelligent. That is precisely why it remains free, governed in the best way, and capable, if it chances upon a single constitution, of ruling all the others.[16]

When done wincing, readers see that not only is ruling others somehow a virtue, for Aristotle, it is a virtue vouchsafed by the type of climate in which a society exists.

What Aristotle is offering, in other words, is a justification for political supremacy on the basis of racial supremacy, which itself is based on climatic supremacy. As he seems to see it, Greece's "intermediate" position between climatic extremes is precisely what makes it capable of enacting political extremes in the form of "ruling all the others." The racist implications should be apparent: climate determines race and race justifies power, even enslavement, making those in more favorable climates racially superior and therefore worthy of political dominion over others.

Such are the dangers of climate determinism, that it doesn't stop at acknowledging the legitimate ways that different climates might in fact lead to different social and cultural formations, including norms, ways of life, and cultural techniques suitable to thriving in one's particular physical environment. Living in the desert, for example, where water is scarce, *does* require different ways of being than, say, living in a temperate rainforest, where water is plentiful. What climate determinists in the Aristotelian vein posit, however, is the superadded element that some ways of life are intrinsically better than others, thereby making a normative valuation instead of what might be merely a descriptive observation about the resourcefulness that different people exhibit to thrive in the characteristic ways demanded by their geophysical circumstances.

While it would be reasonable to interpret climate determinism of this sort as invested in what's only more recently come to be known as differences in "race," that would be insufficient even if it weren't an anachronism. Beyond Aristotle, and more generally across the Greek and Roman traditions that gave rise to the study of climates in the first place, climates have long been held to determine numerous characteristics, demeanors, and abilities of those living in them. Other determinisms would emerge in the millennia to come, spurred by appropriations or misreadings of scientific theories of their own time—for instance, evolution (Darwin), genetics (Mendel), or race as a construct (Boaz), among others. But for ancient Greek and Roman scholars, climates were *the* key organizing principle of human differences. If climatology, then, was not at least a tacitly racist enterprise from the start, it was definitely a differentiating one.[BM]

The very work of identifying and studying climates has always involved dividing the planet into different regions, each with different characteristics, and treating some as more amenable to human flourishing than others. Consider, to take a contemporary example, the popularized identification of those "Blue Zones" across the planet where people have the longest life expectancy. In 2008, Dan Buettner and his collaborators identified five places in different regions around the earth—yet in strikingly similar climates—where residents live longer lives than anywhere else: Okinawa (Japan), Sardinia (Italy), Ikaria (Greece), Nicoya (Costa Rica), Loma Linda (California).[17] Though the research into these regions examined lifestyle tendencies of their residents, such as their social networks, diet, and physical activity, the principle of climatic influence is effectively the same as it's always been. The Blue Zones are largely coastal communities with temperate weather. And, while, race doesn't appear to be a salient demographic

BM This list of disciplines that are complicit in histories of racism is striking, and reminds me of some of the social science disciplines that are also shaped by commitments to determinism, and related practices of differentiation and categorization, too. I'm thinking here of our reading group's engagement with Michelle Murphy's *The Economization of Life* and the way in which economics, demography, and sociology have fed into these racialized, global histories. Of course, rhetoric is on this list too, which raises the question of how our approaches to theorizing, studying, and teaching rhetoric need to change to disrupt these racialized patterns. I see this as a question we address in different, situated, and reflexive ways across this book.

factor, what stands out is that the residents in these places didn't just live long lives; they lived the longest "disability-free" lives. Okinawa in particular thrived in that regard. In chapter 3, Jennifer writes about some of the ways that (digital) social climates foment what she calls "benevolent ableism." If the Blue Zones risk doing something similar, it's based at least tacitly on the idea that geographic climates have an outsized say in what prevents disability to begin with.

If beliefs about the geographic nature of climate determinism linger even today, then they are long-ingrained. In this sense, the major precursor of climatology began a couple hundred years before Aristotle, in the sixth century BCE, with Pythagoras, who first identified Earth as spherical. Doing so made it possible for Parmenides later to expand upon this novel idea and partition our planetary sphere into five zones: two frigid (at the poles), one torrid (around the center), and two temperate zones (in between). Technological developments in the fifth century, such as Anaximander's sundial and gnomon, enabled more precise measurements of the sun's altitude and, hence, of earthly latitudes, but it wasn't until Eratosthenes, born about half a century after Aristotle's death, that climatology had its next major development.[18]

Eratosthenes is most known today for coining the term "geography," but his interest in climates was at the center of this work. Eratosthenes sought to measure the dimensions of the (humanly) habitable regions of the planet. To do so, he delimited the land from what he understood as a surrounding ocean, and partitioned the globe into lines drawn parallel to the equator rather than equidistant to it. From the ancient Greek word *klima*, meaning inclination, slope, or latitude, the very idea of "climates" (*klimata*) comes in large part from Eratosthenes's deduction that these different zones, different latitudes of the planet, have different temperatures depending upon the time of the year, each according to their position relative to the sun. Though Parmenides had reached a version of this insight before, Eratosthenes utilized more precise instruments and techniques of measurement to distinguish latitudinal regions both spatially and temporally. Climates were geographically fixed yet temporally variable. Their absolute position on the planet remained the same, while their relative position in the cosmos changed according to the time of day and season.

Unfortunately, the tidy logic of absolutes and fixity have had a tendency to overshadow the messier logic of relativity and change. Climates can determine the human situation only within a form of reasoning that privileges geographic homogeneity across space and temporal consistency across history. What's interesting is that theories of climate determinism in antiquity seem born as much from scientific or technological advancement as from foreign conquest and the expansion of empire. The Greek and Roman empires, after all, depended on the slave trade, which involved travel to different climates for purposes of plunder more than pilgrimage. This exposure to difference—human and climatic alike—might well have led to its celebration in light of the resilience and resourcefulness that different peoples have exhibited in order to thrive in their own particular ways. Instead, it involved intensely partial perspectivism, whereby the likes of Athens or Rome were held to be the center of the civilized universe, casting everyone and everywhere else as inferior counterpoints on the margins.

Compare, for instance, the Middle Ages, when access to the knowledge of antiquity was largely sequestered and travel to Europe was uncommon for those outside it. Considering that many ancient Greek texts, most famously Aristotle's, were kept alive principally in the Middle East, whose scholars translated it into Arabic at the behest of the Abbasid caliph al-Ma'mūn (786–833), it might seem likely that theories of climate from Muslim geographers expanded upon those of the ancient Greeks. And to a degree they did, just not the way you might expect. The Muslim traditions didn't follow the Greco-Roman tendency to imagine the Mediterranean region that was home to Athens or Greece as the center of the world (and hence those people as the ideal), but they did follow the Greco-Roman tendency to imagine *their own* region as the world's center (and hence their own people as the ideal). Bracketing whether the geographic measurements led to this belief or the belief biased the measurements, Muslim geographers divided the planet into seven climatic zones, as Eratosthenes had, but they located the central and most ideal zone further south, in the fourth zone that spanned across North Africa, Syria, Persia, and Iraq, making Baghdad the center of the world.[19]

The study of geography was at this point largely coterminous with the study of human geography, which was at least tacitly a kind of climatology. In other words, insofar as the work of geography involved studying climates, it already involved the work of studying people. Though the details could be explained differently depending on the scholar, the determinist connection between climates and people was effectively taken for granted, even as it was often used to justify cultural or racial superiority in line with the perspective of the one undertaking the analysis. Even a condensed history of this phenomenon as it continues to the present day can fill a book of its own.[20] Though our aim presently is not historiographical, by showing that discursive rhetoric around climates has always been entangled with rhetoric about race and human hierarchy, it becomes possible to recognize an even bigger proposition: that whatever one understands the polysemous concept of a "climate" to mean, climates seem to exert a material-discursive influence on humanity, certainly in terms of our understanding of ourselves and our planetary condition, but by extension in terms of what it's possible for humans to be and do relative to the imperatives of our earthly coexistence not just with other people, but with more-than-human existents as well.[CR]

Inconsistent Diversity

At least as far back as 1375, the link between climatic variation and human variation was established enough to be historicized by the renowned Muslim historian Ibn Khaldûn, in *The Muqaddimah: An Introduction to History*. For Ibn Khaldûn, the idea that climates determine human diversity was so fundamental as to be a first principle of history itself. The book's first chapter, for instance, opens with prefatory discussions on "the influence of the air upon the color of human beings and upon many other aspects of their condition," and on "the influence of the air (climate) upon human character."[21] Note not just that Ibn Khaldûn understands climate as influencing visible racial difference and character alike, but that he (via his English translator) follows the practice, common at least since Hippocrates but continuing well through the Enlightenment, of understanding "air" to be such a constitutive element of any given "climate" as for the two to be nearly synonymous.

Following the contemporaneous standard of dividing Earth into seven climatic zones, inherited first from Eratosthenes and reinforced later by Ptolemy, Ibn Khaldûn identified the first, second, sixth, and seventh zones as

---

CR I find the splitting and intersecting of the definitions of climate-related words (into the more environmental sense and the more social/political/rhetorical sense) fascinating, too, along with the tension between climate determinism and climate change. These splits/collisions appear everywhere in the genealogy. Climate and obliquity derive from words that mean to lean, to be slantwise, that which slants, leans, inclines everything within a climate to configure in one way or next. Whatever is possible (to be, do, for change) or not, given this, seems a throughline in all our work.

"far from temperate," and hence home to people ("Negroes" and "Slavs") whose "remoteness from being temperate produces in them a disposition and character similar to those of the dumb animals."[22] By contrast, Ibn Khaldûn claimed, the fourth and most central zone accommodated superior people, both physically and behaviorally. "The physique and character of its inhabitants," he writes, "are temperate to the (high) degree necessitated by the composition of the air in which they live."[23] It's not hard to see how this sort of argument can be used to justify the comparative dominance of some people over others by giving it an "air" of inevitability. *Well, well, people can't control the air, of course, climates just are what they are, regardless of any human interventions. That's just the way it is.*[24]

The trouble was that Ibn Khaldûn's own home, in the Arabian Peninsula, fell squarely in the intemperate second zone, making him and his neighbors presumably closer in disposition to the savage animals he disparaged than to the civilized human beings he extolled. Instead of abandoning the first principle that "the air" determined the important differences in humankind, though, he concocted a clever workaround. Despite not actually being within the temperate fourth zone, Ibn Khaldûn claimed that the sea around Arabia cooled it off enough to foster exceptional and more sophisticated people, explaining, "The humidity of (the sea) influences the humidity in the air of (the Arabian Peninsula). This diminishes the dryness and intemperance that (otherwise) the heat would cause."[25]

While the example of Ibn Khaldûn may seem extreme, his is just one among many examples of smart people keen to recognize that indeed there is something "in the air" that seems to influence those it envelops. Importantly, such a view isn't just limited to those, like Ibn Khaldûn or Aristotle, who take the climate to have such a dominant influence on the human condition. Climates need not be comprehensively deterministic to be differentially influential. As Lucian Boia observes in his book on the history of climate determinism, "Nobody, not even the builders of theories (or so we may hope), is so narrow-minded as to claim that there is nothing outside the climate (or collective geographical factors). Human beings, individually or collectively, and history in its entirety, are stimulated by a multitude of factors. Even the most deterministic theories acknowledge this complexity, although they always presuppose a foundation on which to erect the entire theoretical edifice."[26]

The constellation of philosophies associated with climate determinism, that is, are themselves comprised of great diversity and variation, and particularly so when seen in the comparative context of different historical epochs. It seems that the "winds of change" across time, to borrow the hackneyed phrase, are also a kind of climatic influence on what it's possible to think and know within their context. And historical contexts, like regional weather trends, are never entirely uniform or without contradictions. As Boia observes of the Enlightenment, for instance, "On the one hand, the West was increasingly proud of its superiority, a condition that encouraged racism. On the other, it was marked by a universalist spirit that denied any essential superiority."[27] The story of climate theory in the Enlightenment and beyond is too big to tell here, of course, but it passes through the likes of Carl Linneaus, Montesquieu, David Hume, Edward Gibbons, Voltaire, and many other intellectual giants that remain well-known today, though generally not for their deep investments in the race and climate connection.

The best way to illustrate the contradictions and inconsistencies of climate determinism may well be through one of the first people to point them out. Several hundred pages into *The Anatomy of Melancholy* (1621), Robert Burton's long and motley book about what's now called depression, he does so in a curious "Digression on the Air." Burton's digression interrupts the second "partition" of his narrative, on the subject of how to cure melancholy, and so the book's thru-hikers likely encounter it expecting a variation on the fresh-air cure argument. The idea that spending time outdoors in cool, fresh air might cure any number of ailments, from tuberculosis to addiction, has long been put into practice in sanatoriums and other health care facilities around the world. Fresh air treatments have even been part of COVID-19 responses, both to minimize exposure, to stop people from going stir-crazy indoors, and for respiratory reasons once exposed. But Burton mostly just floats around the fresh-air cure. He rather writes about the many inconsistencies in claims, by then commonplace, about a direct relationship between the planet's different regional climates and the types of people living in them.

Burton is gobsmacked by the sheer diversity of life on Earth. All the different types of plants, animals, people, colors, climates, temperaments, and temperatures: it all inspires in him prodigious wonder—and confusion. His interest is less in classifying so much diversity into hierarchies of value, however, than in airing his exasperation over the lack of a universal system capable of explaining it all without running into contradiction. Burton's *Anatomy of Melancholy* was first published in 1621, but already in its fifth edition by his death in 1640. Between the 1621 and 1638 editions, he'd searched widely enough for the "true cause" of so much inconsistent diversity that his "Digression on Air" section had expanded in length by more than half.[28] And still an explanation eluded him:

> Whence proceed that variety of manners, and a distinct character (as it were) to several nations? Some are wise, subtile, witty; others dull, sad and heavy; some big, some little, as Tully de Fato, Plato in Timaeo, Vegetius and Bodine prove at large, *method. cap. 5.* some soft, and some hardy, barbarous, civil, black, dun, white, is it from the air, from the soil, influence of stars, or some other secret cause?[29]

Burton just doesn't know. But of the many dissatisfactory explanations, he shows special skepticism toward the predominant view that the consonance of climate and latitude is to blame. How could it, when, for instance, Moscow is extremely cold year round while Ireland (at effectively the same latitude) has temperate winters? And if the congruence of latitude and climate were such a determining factor in physical appearance, why is it that Spaniards and Italians are "white" while, at the same distance from the equator, those on the Cape of Good Hope are "blackamoors"?

As a wide reader of astronomy, astrology, cosmology, cosmography, geography and more, Burton recognized that Earth was a planet with its own motion, and he seems to have extrapolated from its own unruly plurality that the vast universe itself was a plurality of infinite worlds.[30] Given this sense of infinite plurality as a universal condition, the uniquely wonderful and perplexing diversity of life on Earth could best be explained by what all earthly latitudes and climates had in common: the air. "In all particular provinces," Burton writes, "we see it confirmed by experience, as the air is, so are the inhabitants, dull, heavy, witty, subtle, neat, cleanly, clownish, sick, and sound."[31] Though topography, soil composition, vegetation, and many other factors could contribute to a given atmosphere's distinctive character (perhaps including the cooling effect of Ibn Khaldûn's Arabian Peninsula), the air was the most subtle and essential one of all, for it was the air that most depended on the climatic circumstances. "Whether hot, cold, dry or wet," Boia summarizes, "the air is the medium through which the climate transmits its messages. . . . Wherever we are, we breathe the air specific to any locality and behave accordingly."[32]

Convenient Falsehoods

Today, it's easy enough to see that climate determinism has often rested on many convenient falsehoods that have helped to bolster the personal interests of those making them. It's no accident that, for Aristotle, Athens had the ideal climate and people. For Ibn Khaldûn, Baghdad. For Jean Bodin (whom we didn't discuss), it was Paris. Even Burton—despite his suspicions about climatic influence on human difference—needed *something* to be instrumental to human disposition. How else to offer a cure to melancholy? So, he settled on the air and ultimately urged people to get outside where the air was fresh. This sort of move-the-goalpost thinking is certainly not unique to old or new discussions about climate determinism and the racism it often exists to justify. But these days convenient falsehoods about the influence of climates on people also need to face the reciprocal "inconvenient truth" that people influence climates.

If the contemporary geological epoch now often known as the Anthropocene is fraught with inconsistencies of its own, as Burton would surely argue today (alongside many others who already have),[33] one of these inconsistencies is to risk implying that the Anthropos, humanity, is an undifferentiated whole. By positing a new geological epoch in the planet's history marked principally by the human impact on its fossil record, the idea of the Anthropocene suggests that collectively, not individually, humans have determined the planet's material history, including by changing its global climate. This is quite a different proposition than the old version of the story, whereby earthly climates determine the fate of people. The issue isn't just that humans and their constructed byproducts have been around long enough already to have become fossilized, though the ruins of prehistorical civilizations are indeed well buried. To identify a new and current geological epoch following the Holocene is also to project an anticipatory vision of Earth: that is, a speculative view of what the planet will eventually be understood to have been from the standpoint of geologists thousands of years from now.

From such a standpoint, it's likely that the record will disclose a story of global industrialization, built on rapid human exploitation of planetary resources until ecosystems collapsed into a major extinction event. Such a story, told thousands of years hence (if people are still around to hear it), likely won't be nuanced enough to identify specific actors or agents of the destruction. It will be humanity, as a whole, that was the scourge on planetary coexistence. Despite the many who are trying, telling such a story today, even as a cautionary tale, is terribly hard to do in face of the deeply entrenched tendency to imagine the individual human, possessed with free will and motive, as the quintessential agent of change. Such character types abound in our myths and stories: the hero, the genius, the rebel, the leader, the champion, the warrior. Whatever implicit bias about gender or race may lurk in these stereotypes, a stubborn belief in each person's own individual capacity to act freely, make decisions, and instigate change is hard to reconcile with the geological long view, in which none of anyone's individual actions, however significant, will even register as a minor influence relative to their envelopment within the collective whole of humanity.[JL]

Yet, the very story of climate determinism, as told here as elsewhere, nevertheless does implicate individual actors who have had a disproportionate influence on the idea taking hold—and therefore its being used to justify appropriative and violent acts. The apotheosis of such a self-justifying worldview probably came in the late nineteenth century, when determinist thinking scaled up to serve the expansion of empires through settler colonialism based on a new premise, this time wrapped in a post-Darwinian whiff of the scientific. Friedrich Ratzel, a renowned fin-de-siècle German geographer, introduced the idea of Lebensraum—or biogeography—in a review of Darwin's *Origin of Species*. Woodruff Smith describes Ratzel's sense of Lebensraum as "the geographical surface area required to support a living species at its current population size and mode of existence."[34] But the key implication was that all species, including humans, adapt in accordance with their shifting Lebensraum, that is, by expanding their reach, when necessary, into new geographical territories through migration and colonialization—each being justifiable on the suspect basis of that being what all species do to survive.

Lebensraum was accordingly yoked to a determinist logic of a state's manifest destiny. We all know what happened when Germany tried that out in the twentieth century. But it doesn't look much different in a British, American, or any other settler colonialist context. If nineteenth century notions of Lebensraum represent an apotheosis of the old climate determinism fallacy,

---

JL I don't think the lens of rhetorical climates fixes this conundrum, but it does help push back against any nihilism-tinged responses in the emphasis on that which is unavoidably shared. Perhaps, to pick up on Nathan's earlier point, we need to develop more modes of rhetorical sensing that encourage looking for shared surfaces. Or, where are the freckles in our eyes aligned?

however, there is another fallacy that has since emerged in the pendulum swing away from it. Mike Hulme calls it "climate indeterminism," in which climate is wholly invalidated of any explanatory power when it comes to human physical traits or cultural tendencies.[35] Anthropocenic thought can tend toward such a position. In place of climates determining human traits and behavior (and therefore being used to justify exceptionalisms), in the indeterminist view, climates themselves are relegated merely to a passive byproduct: climates don't shape human character; human character shapes climates.[NS]

And it's here where our airborne digression has finally grown tall enough to reach the ground. After all, another character type in at least Western myths and stories—certainly those long-told about democracy and the public role of citizens in shaping society—is the rhetor: the "good man speaking well," whose rhetorical eloquence leads others to assent to action capable of changing worlds. *Non ducor, duco*, says the rhetor: "I am not led, I lead." Climate determinism may always have been a bogus project of projecting and protecting power, but that does not mean that the inverse problem of humans wholly determining the climate's fate is any more accurate. As there's no separating the dancer from the dance, there's no separating change from the conditions amenable to its flourishing. Individuals can act with influence on others, but they are always already acting within a larger enveloping context. Call it a milieu, a structure of feeling, or something else. We call it a climate. Either way, the ambient, material-discursive conditions in which change occurs are just "as rhetorical" as the material-discursive acts that sometimes initiate such change. Each determines the other, intra-actively. Rhetoric is climatic; climates are rhetorical.[CR]

The paradox of the Anthropocene hypothesis is that humans, collectively, have the power to change Earth's climate, while at the same time, not one

---

NS I find this history so informative and thought provoking. I am particularly struck by how climate has been about the human, defining the human as a function of place and condition, so much so that it is strange that all the many critiques of humanism have not directly addressed it all that much. And it raises questions about how transcendent visions of humanity articulate themselves with and against climate determinism. There are many synonyms, but as a "keyword" climate is powerful and weighted down with its history.

CR This makes me wonder how change is even imaginable, let alone possible, from a climate perspective. How might we engage in climate-changing labor, as in the work required to change a climate (presumably for the better, toward justice, sustainable futures, whatever this might mean, etc.)?

of us, individually, has any power over the changes that began before us and will continue after. Of course, anticipatory speculation aside, in this moment, it's far from the case that all people leave an equally damaging "footprint" on the planet. What we know about climate change at a planetary scale is that its materialization is less uniform and steady across time and space than it is patchy and nonlinear. Humanity is not an undifferentiated whole, and the destruction of our planet is as unevenly enacted across human geographies as it is unevenly impactful on human populations. It's well known that the biggest nations and moneyed interests contribute most to rising climates, while the smallest and poorest suffer most from its effects. As Kathryn Yusoff puts it, "To be included in the 'we' of the Anthropocene is to be silenced by a claim to universalism that fails to notice its subjugations, taking part in a planetary condition in *which no part* was accorded in terms of subjectivity. The supposed 'we' further legitimates and justifies the racialized inequalities that are bound up in social geologies."[36]

People have refused this story before, back when climates were held to determine the human condition. Now, when humans are held to determine the climate's condition, it is essential to recognize that these climates are both meteorological and social, that the two cannot viably be separated—indeed, that they have always been entangled, such that climates influence people and people influence climates. But recognizing as much isn't enough. To the contrary, it compels further work to be done, such that the remit of those studying rhetoric is neither to promote the anthropocentric vanity of imagining that humans can fix anything, nor to loose anarchy on the world when the human center will not hold. This will mean paying closer and more critical attention to the ways rhetoric can operate as a dispensation to determine the "we," whether as a product of human intention or not, with the consequence of fostering different kinds of inclusion and exclusion at the expense of others. Of course, these climatic energies are terribly hard to see, but that's because they're so often taken for granted either as inevitable or reliable. In the next chapter, we'll see how this taken for grantedness manifests at the level of a body's flesh, where Whiteness itself acts as an injurious climate, widely taken for granted because it disavows any refusal to grant what Whiteness has taken.[CI]

CI Oh, and about that joke. *I'll tell you later.*

Microclimate 1 (Nate and Chris)

Chris's opening chapter calls one to reflect about the terminological habits of rhetoric on offer in the title of our book, because the "modifier move" so common to studies of rhetoric—rhetorical X and X-ical rhetoric—is illuminated unexpectedly by its application to climate, particularly by the ancient logic of climatic determinacy. One can see this logic on display when rhetoric is made the climate of other things and when those things are made the climate for rhetoric. Ethical rhetoric—rhetorical ethics. Violent rhetoric—rhetorical violence. Political rhetoric—rhetorical politics. Ecological rhetoric—rhetorical ecologies. Critical rhetoric—rhetorical critique. Philosophical rhetoric—rhetorical philosophy. And so on. In trying to diagnose the subject "rhetorical climatology" by giving it historicity, the very naming of that subject is rendered a symptom of something greater that needs inspection, thus putting the diagnosis under its own analysis. In every instance of rhetoric's possessive modification, including modifying climate with rhetoric and rhetoric with climate, it is worth following the inclination of any such pairing toward what cannot be possessed, where possession falls away. My guess is such a practice will routinely spill over into uncontainable questions about what it is to be human and the forces that that kind of being tries to muster in cooperation or conflict with the greater forces that make its being and non-being possible. The human as the focal point for explaining more than what is human, what is inhuman, and all the relationalities thereof will haunt each and every one.

Reflexively, then, in thinking of rhetoric and climate together, to follow the inclination of the pairing to places where the climatic determinacy of the "modifier move" loses its grip, we might hear these essays as attempting to speak twinned, chiastic premises: *climates animate rhetoric* and *rhetoric animates climates.* This puts one on various paths that lead away from the title, not toward it, by throwing one into a loop that cycles between ancient discourses about Man in wider fields of dynamic contingency, social and biophysical, while unraveling those discourses at the same time. That is maybe some of the challenge here with engaging climate practically and conceptually as rhetoricians: working from some sense of rhetoric as a field while actively opening it to forces beyond its control where it loses its coherence. If rhetoric and climate animate each other, it may be because neither is a distinct category of animacy but are in fact different ways to account for the nexus of forces that animate being and non-being—as agentic yet subject, changeable yet stable, separate but enmeshed. The scope and prospects for livability of each nexus is not something either term can possess, separately or in tandem.

And, of course, different kinds of possession are fundamentally what's at stake. If the very grammar of English requires possession as a rule, the very rhetoric of bodies and their emplacement in particular worlds at particular times also entails an inevitable possession, a being-in of possessedness. "We are condemned to being-in," Peter Sloterdijk writes, "even if the containers and atmospheres in which we are forced to surround ourselves can no longer be taken for granted as being good in nature."[1] To imagine climate determinism or racist climates as rhetorical forces is to ask what sorts of being-in are worth cultivating—and to challenge our condemnation by cultivating better ones ourselves. To say that Whiteness is anti-Black, as Nate does in the chapter ahead, suggests that Whiteness acts as a form of possession that can't be escaped—even when "it" isn't "acting" volitionally. Whiteness is the possession affronted by anything that denies deference to it. Its built-in normativity forecloses mutual flourishing by possessing everything else with comparative shortcomings. The work of dispossession, of refusing this normativity, begins then not just with identifying the *larger* being-in to which we're invariably condemned, but with identifying the *smaller*, mundane barbs and harms that create it.

Consider what microclimates and microaggressions have in common. "I have several Black friends." "Where are you from?" "The most qualified person should get the job." To assume the cultural and social privileges of Whiteness, even in minor and quotidian keys, is to cultivate a White monoculture that, through its possessive precepts, invalidates diversity in advance. And though they're often undertaken with "innocent" intentions, the damage they do is no less consequential. How can the verticality of a feeling, that condensed intensity with which it bores down into you, also have a horizontal aeration, an elsewhere for others? Just as Rachel Carson showed, long ago, that all bodies are permeable, that poison flows downstream and seeps inside, so too moves social toxicity, finding its way in, but also accumulating like a cloud that becomes its own being-in. Microclimates and microaggressions alike, in other words, can each scale up to create inhospitable macroclimates that don't allow for mutual flourishing because they never cultivate the conditions amenable to it taking hold—on the ground, in the air, in the rhetorical sociality of being commonly alive. And while each oblivious microaggression or presumption of normativity might leave its own material-discursive scar, the math doesn't add up without a lost remainder. A single microaggression can spoil an afternoon, but it can also fill a workplace or a school like a gas filling a room. Such aggressions, however minor, are microclimates in a social key, worlds unto themselves, even as they arise from within those broader climates that they collectively contribute to creating. Unlivability is the ultimate result.

2 Inclement Weather (Nate)

We "do not know what 'whiteness' means" because "there is, in fact, no white community."[1]

Rather, Whiteness is an adaptive arrangement that sustains a world for people who "think that they are white" at the cost, James Baldwin warned, of bringing humanity "to the edge of oblivion."[2] Every passing day Whiteness refashions the "world," producing and absorbing otherness subtly or ruthlessly as it binds impermanent being to the racialized skin,[CI] "blood," and genes of bodies.[3] It is remarkably resilient, yet it inspires fearful fragility in those who identify with it. Whiteness works the remotest corners of life and encircles the planet unremittingly, driving ways of being often too small or too large to notice.

CI And what's skin if not a surface? So, surfaces again, but scaled up here to a climate of "Whiteness" made normative as a superficial appearance that gets held to indicate some supposedly interior, sub-surface, "deeper" qualities, such as blood or genes. My hunch is that rhetoricians thinking about surfaces might point to Aristotle's writing about the power behind the "appearance" of things via his concept of *phantasia* (e.g., see Ned O'Gorman, "Aristotle's 'Phantasia' in the 'Rhetoric': 'Lexis,' Appearance, and the Epideictic Function of Discourse," *Philosophy and Rhetoric* 38, no. 1 [2005]: 16–40; or Debra Hawhee, *Rhetoric in Tooth and Claw* [Chicago: University of Chicago Press, 2017]). Ancient studies of rhetoric in general were pretty fixated on appearances, including at the level of the body (see, e.g., Jens Kjeldsen, "Talking to the Eye: Visuality in Ancient Rhetoric," *Word & Image* 19, no. 3 [July–September 2003]: 133–37). But anthropologists and cultural geographers interested in surfaces today can make that rhetorical work seem—wait for it—superficial. Check out, for instance, Joseph A. Amato, *Surfaces: A History* (Berkeley: University of California Press, 2013); or Tim Ingold, *Lines: A Brief History* (London: Routledge, 2007). Both point to the ways surfaces run deep enough on their own. Rebecca Coleman and Liz Oakley-Brown's veritable inauguration of "surface studies" in their special section of *Theory Culture and Society* makes this point as well as anything I've seen. See Rebecca Coleman and Liz Oakley-Brown, "Visualizing Surfaces, Surfacing Vision: Introduction," special section, *Theory, Culture & Society* 34, nos. 7–8 (2017): 5–27. Start with surfaces. Climates will follow.

I discuss Whiteness as purportedly the most *human* human being—as bleach for Europeanized imaginaries of peak, deserving humanity naturally sitting atop nature—not as communities or identities. Chauvinist and fragmented, Whiteness is oppositional: normatively global, patriarchal, Occidental, anti-Indigenous, anti-Semitic, able.[4] Of course, it is certainly also anti-Black. That's my first, wholly unoriginal, claim: Whiteness is inseparable from Blackness and endemically anti-Black. In that regard, I discuss Blackness not as communities or identities either. Following Jared Sexton, I view Blackness as critical to raciality without being a synecdoche for race: "Black existence does not represent the total reality of the racial formation—it is not the beginning and end of the story—but it does relate to the totality; it indicates the (repressed) truth of the political and economic system."[5] Anti-Blackness is resilient as Whiteness is resilient, changing constantly yet maddeningly constant, and although characteristic of coloniality, anti-Blackness exceeds Whiteness (another unoriginal claim).[6] As Eric King Watts argues, "structural antagonisms are not static, but are movable and moving configurations."[7]

Writing this, I am motivated by the exigent need to account for Whiteness, which scholars of color and allies have repeatedly demonstrated is constitutive of the study of rhetoric *and* by the simultaneous need to account for anti-Blackness, which Black scholars, writers, and artists have repeatedly demonstrated is constitutive of Whiteness. At first glance, this seems dyadic, recentering Whiteness against Blackness, rather than thinking triadically perhaps ("European-Negro-Indian") as Sylvia Wynter advocates, or figuratively as Anne Anlin Cheng does regarding "Asian/Asian American women in American culture" in *Ornamentalism*, or intersectionally as Kimberlé Crenshaw and many others have argued for.[8] However, I am intent on accounting for anti-Blackness as *decentralized* within a Whitened environment, not dyadically, such that it variably affects power configurations. I ask, how to think about Whiteness in its anti-Blackness and in terms of rhetoricity if antagonisms are racialized and highly configurable? How to talk about race, rhetoric, and anti-Blackness without making White identity formation against otherness primary, or blurring dispossessive, territorializing projects and their violence into one another, or perhaps worse, establishing hierarchical "taxonomies of violence"?[9] The answers are not clear, although many have discussed these questions and given much to weigh in response.[10]

I don't have answers either (it would be galling to think I did), but for my part I want to return to Thomas Nakayama and Robert Krizek, who, echoing Baldwin and anticipating arguments such as Watts's, explain that Whiteness has no set form.[JA] In "Whiteness as Strategic Rhetoric," they describe Whiteness as a territorializing assemblage, arguing that Whiteness "is not itself a place, but it functions to resecure the center, the place, for whites." I underscore their characterization of Whiteness as a "milieu present at the intersection of differing 'realities'" because it describes more than Whiteness as an invisible center with privileges arrayed to protect its centrality.[11] Indeed, Nakayama recently noted that "*whiteness or Whiteness is not contained.* Its cultural logic is expressed and performed around the world in ways that are not unexpected. Whiteness is not a national phenomenon but an international one."[12]

I believe a concept of climate can extend Nakayama and Krizek's foundational work and aid understanding of anti-Blackness as a decentralized dimension of Whiteness. The term *racial climate* has been around for some time, but Christina Sharpe's description of anti-Blackness as climate informs my thinking and I place her in dialogue with recent work from Lisa Flores and Megan Eatman.[13] In brief, you can expect an anti-Black climate from Whiteness; violence, injustice, and nullification are its characteristic weather.[CI] Thinking of a milieu climatically looks past racism as intersubjective assault (the "perpetrator perspective" Darrell Wanzer-Serrano calls it) to the anti-Black worldmaking of Whiteness and to the modes of fugitivity that refuse it.[14] Anti-Blackness is irreducible intersubjectively because it is an omnipresent condition of antagonism in a Whitened world.

I first describe anti-Blackness and Whiteness ecologically in relation to rhetoric. Primarily I turn to Black feminist scholars to magnify Whiteness as a milieu, also noting the value of Thomas Rickert's concept of ambience

---

JA Ummmm, I want to say that Whiteness controls all the forms, through the settlement, the metropole, the wall, the line (see CI's previous note from this chapter), the prison, the plantation prefiguring the city. Nate, you bring much precision to the distributed non-place of whiteness (lowercase) . . . I just see its it-ness in buildings, surveillance devices, the map that obscures the territories, perhaps the milieu to come.

CI Graduate student readers, an assignment: Expand this sentence into a dissertation. Ready? Go.

for describing the pervasive anti-Blackness of Whiteness. With viewpoint established, I qualify an ecological conception of Whiteness by comparing and contrasting Sharpe's "anti-Blackness as a total climate" with Flores's analysis of rhetorical climates of deportability and disposability and Eatman's elaboration on Sharpe through "ecologies of harm." I close by reflecting on incommensurable relations within an anti-Black climate and Whiteness as an anti-Black meta-ecology.

The Anti-Blackness of Whiteness

Nakayama and Krizek's "milieu" resonates with a long history of characterizing Whiteness as pervasive, dynamic, and environmental such as Édouard Glissant describing race as a fusion of environment and consciousness, not to mention scholars who also turn to assemblage theory such as Alexander Wehiliye and Achilles Mbembe.[15] The range of scenic, systemic, encompassing descriptions of Whiteness is extensive. For example, Pugilese writes that Whiteness "so constitutes the molecular fabric of everyday technologies and practices that it cannot appear as a category as such. It is precisely because whiteness is so inscribed into the material weave of everyday life that one cannot talk of whiteness as such."[JA] Because Whiteness is never "essentially identical to itself," it is infrastructural and active.[16] To the extent that Whiteness traces a color line that pervades and organizes sociality as Du Bois explained; that it haunts Black and Indigenous life as unceasing violence; is concomitant with European coloniality; is a metaphysical and psychical condition; is contingent on Blackness as an ontological state of non-being; is fundamental to American racial grammar; is a material factor in ecological demise; is a globalized, racial theodicy; is a global matrix of fear; is an operative condition of Western modernity . . . then Whiteness is atmospheric.[17]

Conceiving Whiteness as an unavoidably anti-Black milieu risks reducing complex structures (those intersecting realities) to expressions of anti-Blackness alone. Broadly speaking, Whiteness produces invisibility, dispossession, and exclusion and can make racial dominance, violence, and injustice unrecognizable even when racial power is nakedly exercised. The gaslighting of Whiteness is well

JA Okay, here we go. Thanks, and whew! The "molecular fabric of everyday technologies" for sure.

marked: you did not suffer the violence you suffered; you did not see the violence you saw. How then to qualify the rhetoricity and anti-Black character of a milieu that does more than suffocate Black life?[18] Anti-Blackness is not conceivable without Whiteness, but neither is it redundant nor only a sub-component. I suggest considering Whiteness as a worldmaking milieu, or meta-ecology, that constrains the co-invention of "humanity" and "world."

Black feminist thought is vital to understanding the necessary but contingent relations between anti-Blackness and Whitened worldmaking, and I turn to the work of Denise Farreira da Silva, Saidiya Hartman, Zakkiyah Iman Jackson, and Katherine McKittrick specifically. In *Toward a Global Idea of Race*, Silva diagnoses what she names modernity's "analytic of raciality," which legitimates a global order of colonial dominance and expansion. The analytic was forged, she argues, through racial science and philosophies of history that posited raciality as pre-existent to being, such that different races had unequal capacities of reason and susceptibility to external conditions, which manifested themselves in space and through time. Whiteness in the form of *homo modernus* was defined not only against Africanity but against disparate racialized subalterns, which she calls the "affectable others of Europe," who, unlike Europeans, were considered unable to transcend their ontic limits (that is, were determined by their circumstances). The purported representativeness of Whiteness to the human condition and the non-representativeness of affectable others has everything to do with Man supposedly occupying a privileged relation to Reason and the power to self-create, with non-White others being defined by their ontic limitations.[19] Noting the ableism intrinsic to the self-sovereign *homo modernus*, Rachel Gorman writes that "affectable other must be known, but cannot be self-knowing."[20] Relative to other subalterns, the affectability of Blackened being was/is cast as uniquely adaptable to brutal conditions but incapable of reason and distinctly other to Whitened being. Saidiya Hartman described this adaptability in the context of chattel slavery as the fungibility of the commodity, which "makes the captive body an abstract and empty vessel vulnerable to the projection of others' feelings, ideas, desires, and values." In slippery opposition to the indivisible self-creating sovereignty afforded to Whiteness, fungibility names the "elasticity of blackness and its capacious affects" that can serve as "a vehicle for white self-exploration, renunciation and enjoyment."[21]

In *Becoming Human*, Jackson goes further, building from Hartman to theorize the fungible ontology of Blackness as,

> plasticity, which maintains black(ened) people are not so much as dehumanized or cast as liminal humans nor are black(ened) people framed as animal-like or machine-like but are cast as sub, supra, and human simultaneously and in a manner that puts being in peril because the operations of simultaneously being everything and nothing—human, animal, machine, for instance—constructs black(ened) humanity as the privation and exorbitance of form.[22]

Like Hartman, Jackson stresses that producing the plasticity of Blackened humanity, and indeed the productivity of that plasticity, is paid for by the "physiologically and psychologically debilitating effects of antiblackness," which are immeasurable.[23] Anti-Blackness not only presupposes that inherent disability defines Blackened being—which has been exceedingly effective for construing fearful visions of humanity and its others—but actively produces debility through what Yanar Hashlamon, building from Jasbir Puar, describes as "making available, or capacitating, certain rhetoricities for some populations at the expense of others."[24]

If Blackness is an integral but threateningly plastic limit for Whitened humanity, then anti-Blackness is a disastrously capacitating force of raciality that calls for a more versatile relational grammar than margins and centers. In *Demonic Grounds*, McKittrick approaches this problem through human geography. There she explains that language about raced and gendered cultural peripheries incompletely addresses the changeable, formative materiality of anti-Black experience because "margins and invisibility . . . are also lived and right in the middle of our historically present landscape." Her argument amplifies the meta-ecology that Silva recognizes in a global analytic of raciality and that Jackson does in Blackened plasticity. To the extent that affectable others of Europe are spatially, temporally, and corporeally dynamic subject/object positions, they enact geographies which McKittrick defines as "space, place, and location in their physical materiality and imaginative configurations."[25] For McKittrick, who discusses Black life in diaspora, "Black geographies comprise philosophical, material, imaginary, and representational trajectories; each of these trajectories, while interlocking, is also indicative of multiscalar processes, which impact upon and organize the everyday."[26] Black geographies are necessarily shaped by the costly plasticity of Blackness: "those who continue to inhabit the uninhabitable are so perversely outside the Western bourgeois conception of what it means to be human that their geographies are rendered—or come to be—inhuman, dead, and dying."[27] Yet Black geographies also disrupt the analytic of raciality because they extend beyond actions of Blackened people, distributions of bodies, and normative cultural locations; in that sense, Blackness is a host of paths and flights that traverse being, radiating from Black experience and the refusal to succumb to anti-Blackness, not from a margin.

It follows that Whiteness is not a center but ambient in Thomas Rickert's terms in that it "permeates each aspect of the emergent ecology"; or, as Nakayama and Krizek wrote, the "intersection of differing 'realities'" is not found on a map.[28] The centering effect of Whiteness is what the milieu feels like as it territorializes "other" geographies whenever and wherever trajectories converge. Whiteness-as-milieu makes anti-Blackness endemic to Western European attempts to invent a Whitened world where everything is enclosed within linear, hierarchical, universal logics (what Silva calls the "scene of engulfment").[29] Whiteness thus is a not only a strategic rhetoric that secures White supremacy but also an ambient one.[CR] More than that, putting Rickert in conversation with Jackson, the anti-Black ambience of Whiteness emanates from "an ecology of violence—pervasive and chronic" that is intrinsic to European racialized order.[30] Over and above acts of violence and dispossession, anti-Blackness permeates raciality through the "material affectability" of violence that "knits itself into the environment"[JA]—a feeling for what the middle of the historical present can and cannot be.[31] From this point of view, Whiteness is not only

---

CR This stretch on Whiteness-as-milieu reminds me of work on linguistic justice and antiracist praxes in writing studies. Deep into these ideas in that context, it's now so transparent to me how racist language ideologies tacitly reinforce White cultural and linguistic norms as the gold standard within language and writing pedagogies to the detriment of BIPOC, working class, multilingual, and other marginalized communities. But these comings together of language ideology, racism, and colonialism are atmospheric in just the sense you are talking about, Nathan. This coming together becomes part of the climate, naturalized, like the air we breathe, invisible, tucked in places we've forgotten, like standard dominant English as the unchecked default in the classroom/public or in ways we make unconscious assumptions about people based on their language use.

JA I wish I saw this, said this, or something like this just to let the words fumble about, but I've certainly dreamed this weather as a deepening, grounding, particulate, excessively filmy ambience. So cool.

a normative category, wherein anti-Blackness is discourse used to racialize different peoples as threatening, naturally impaired, and limited demi-humans. Understood climatically, anti-Blackness is an enveloping, visceral attunement to Whiteness that permeates and exceeds civil life, capacitating different rhetorics as the terror and terroir of "impossibility itself."[BM] By "terror and terroir of impossibility," I refer to violent ecologies sustained by efforts to globalize Whitened self-sovereignty, suffused by a plastic, Blackened otherness that is not (yet still is, impossibly) human.[32]

Ecologies Have Climate

In addition to Rickert's highly useful conception of ambience, there is a wealth of ecological views of rhetoric with which to ramify Nakayama and Krizek's description of Whiteness as a milieu. I find Sharpe's discussion of climate to be invaluable for understanding anti-Blackness as an attunement to a Whitened milieu (rather than as a barricade erected at its periphery). She offers a particularly ambient way of thinking with her book, *In the Wake*, where she writes "the weather is the totality of our environments; the weather is the total climate; and that climate is antiblack."[33] For her, climate shifts emphasis away from iterated performances, enacted identities, and embodied locations to the affective force of being engulfed. Drawing on Frantz Fanon, Sharpe explains "it is not the specifics of any one event or set of events that are endlessly repeatable and repeated, but the totality of the environments in which we struggle; the machines in which we live; what I am calling the weather."[34] The weather of an anti-Black climate swirls around the "intersection of differing 'realities,'" not just when violence rains down.[35] To wit, she writes that "antiblackness is pervasive as climate. The weather necessitates changeability and improvisation:

BM I'm noticing this relation between terroir, territorialization, and terror and wondering about how territorialization feeds into the affective condition of terror. The movements across these terms also reminds me of Jennifer Scappettone's discussion of Ralph Ellison's interpretation of Bessie Smith's song "Work House Blues" and how Smith's lines "'Goin' to the nation, goin' to the terr'tor, / . . . / Bound for the nation, bound for the terr'tor' sing the 'terror' in that delimited domain as well as the promise of freedom." Jennifer Scappetone, "Precarity Shared: Breathing as Tactic in Air's Uneven Commons," in *Precarity Shared: Breathing as Tactic in Air's Uneven Commons*, ed. Myung Mi Kim and Cristanne Miller (Albany: State University of New York Press, 2018), 46.

it is the atmospheric condition of time and place; it produces new ecologies."[36] Labyrinthine ecological niches comprise a totality wherein anti-Blackness is anticipated everywhere as inclement weather when realities clash in a Whitened milieu.

Sharpe's core desire it to keep "breath in the Black body" in the face of a total anti-Black climate, which she argues requires an orientation to the wake (of the slave ship, of a vigil, of being wakeful): "to occupy and be occupied by the continuous and changing present of slavery's as yet unresolved unfolding."[37] What she calls *wake work* means "attending to Black life and Black suffering" in its specificity; it is about "grief and memory."[38] Wake work offers a kind of counter-attunement but it differs from cultivating public grievability for precarious life, because, she asks, "how does one mourn the interminable event"?[39] I am reminded of Mark McPhail's pessimism about racial reconciliation and Sharpe adds perspective: the mourning necessary for reconciliation is impossible while ruinous tempests rage on.[40]

Where Sharpe accentuates climate analytically, climate is often implicit in scholarship on rhetoric, but Flores and Eatman have recently authored excellent studies that make climate an explicit concern. Their works strongly resonate yet contrast with Sharpe such that reading them together deepens understanding of anti-Blackness and worldmaking in a Whitened milieu. Flores offers, I believe, the most extensive analysis to date of racial rhetoric in terms of climate in *Deportable and Disposable*. In *Ecologies of Harm*, Eatman extends Sharpe's thinking to focus on the rhetoric of violence generally. E. Cram's recent *Violent Inheritance* also offers important insights into thinking about Whiteness in terms of climate and energy and Ellen Gorsevski's *Peaceful Persuasion* understands climate as a "living, moving, prime 'actor' in unfolding events," but I limit myself because the latter works open further questions that I do not have space to address.[NS]

NS Cram's book was published in late spring 2022, so there was not time to really integrate their subtle and careful use of climate into this chapter or to read it as a group. Also, echoing Chris's comment in the next chapter, we erred in not accounting for Ellen Gorsevski's *Peaceful Persuasion*, which I only learned of late in the eleventh hour of this project. Her seventh chapter lays out a sense of rhetorical climate that anticipates the materialist and ecological thinking that informs this volume: "In the familiar Burkean pentad, the experiential aspect of 'scene' becomes a living, moving, prime 'actor' in unfolding events. This acting organism is a rhetorical climate" (125). Regardless, both these works do open a series of questions that deserve more time and focus than pretending they are simply redundant with what Sharpe, Flores, and Eatman argue. See Ellen W. Gorsevski, *Peaceful Persuasion: The Geopolitics of*

Also, I will not engage the carefully researched and written cases that are the backbones of Flores's and Eatman's books and instead touch on their theoretical contributions as I think they may relate to Sharpe.

Climate as Expectancy

Sharpe and Flores align in using climate to characterize expectations of racial violence based on amalgamated experience, which is consistent with a meteorological view of climate as expected weather.[JL] Climate describes what is probable calculated from observation.[41] Both Sharpe and Flores are concerned with how raciality creates a hostile climate in which humanity and its world are differentially invented, but they vary in the way they conceive worldmaking relative to expectancy.

Through deep examination of historical cases, Flores describes "*rhetorical climates of deportability and disposability*, or those constellations of discourses, cultural practices, laws, and policies that coalesce to produce and maintain constitutive spaces, deportability and disposability frame and solidify Mexicanness."[42] Deportability, she explains, is made "of a multitude of practices, perspectives, and procedures that enable both literal and figural surveillance and regulation" that "produces suspect bodies and lives lived in the 'shadows.'" And, if "deportability is widely theorized as the condition of being liable to deportation," then "disposability refers to the condition of being used and then subject to disposal—lives discarded."[43] She understands "rhetorical climate" to be about "meanings and identities" but is especially concerned with senses of fear and threat that animate the systemic conditions of racial violence and

Nonviolent Rhetoric (Albany: State University of New York Press, 2004); Lisa A. Flores, *Deportable and Disposable: Public Rhetoric and the Making of the "Illegal" Immigrant* (University Park: Penn State University Press, 2020); Megan Eatman, *Ecologies of Harm* (Columbus: Ohio State University Press, 2020).

JL This understanding of racial violence as climactic also dovetails with discussions of microaggressions and how their destructive force comes through the aggregation over time. It's hard to mark why it feels so bad when colleagues repeatedly mistake you for the "other Asian faculty member in economics" without a framework that accounts for amalgamation as rhetorical.

the performative racial scripts and logics enacted by those constellations of discourses, practice, laws, and policies.[44]

Scripts, logics, and affects operationalize expectancy such that Flores describes a kind of atmospheric repertory for rhetorical outcomes—the conditions of deportability and disposability—that emerge from histories of violence, immigration law, migrant worker policies, and visual and linguistic tropes of illegality and alienness. *Climate* encompasses strategic discursivities that foment the identities stabilizing always fragile White hegemony and Mexican American subjection. For example, detailing racial scripts and logics, Flores shows that what she calls "Blackened violence" and "Whitened violence" operated intersectionally during the Los Angeles riots of the 1940s, when Mexican American youths wearing zoot suits were brutalized in a context of war and White national fervor. She explains that the threat of "Blackened violence [attributed to zoot suiters] invokes scripts of racial otherness, already articulated in, on, and through the zoot suit" and "calls forth racial scripts of alienness, here named in the familiar tales of the mobility of undesirable immigrants, those primitive uncontainable hordes that descend with the force of nature."[45] In this way, a fear of "Blackened violence turned on both the body logics of race and the mobility logics of borders, which together layered race, gender, sexuality, and foreignness on Mexican American youth."[46] In response, Whitened violence toward zoot suiters "restored order through the literal disposal of zoot suits and the figural disposal of race."[47] Anti-Blackness informed the violence—Blackened expectations layered the violent threat of Mexican alienness as Whitened expectations of "necessary" violence became "justified, almost innocent."[48] On these terms, an anti-Black climate conditions the prosecution and assessment of racial violence to fit a Whitened world, Blackening bodies and ordering events where racial disposal is anticipated and required.

In comparison, Sharpe approaches climate-as-expectation through her concept of the weather, which names the relentless, variable eventualization of anti-Blackness. She backgrounds the performative aspects of texts, policies, and practices, while still insisting on their importance, to highlight the need to understand "the terror visited on Black life and the ways we inhabit it, are inhabited by it, and refuse it."[49] Doing so stresses that the cumulative affect of anti-Blackness drives ways of being relative to death's specificity, whether death comes fast or slow. Or, rather, that the ambient force of anti-Blackness, weathered at the level of persons and communities, attunes one to the ongoing calamities affecting Blackened people trying to live within Whiteness as the milieu of worldmaking. Sharpe turns climatic expectancy toward what it means to "inhabit that Fanonian 'zone of non-being,'" which is dwelling with "the sense and awareness of precarity; the precarities of the afterlives of slavery."[50] That precarity is the feeling of an ever-present "horizon of death," Gorman argues, a climate of racialized disablement against which the Whitened transparent "I" fashions representations of its self-overcoming.[51] In that sense, a Blackened zone of non-being encompasses not only "the materiality of racialized violence" but also normate visions of Whitened being, which spatialize non-being as inescapable bondage in disablement, the ground of un-life.[52]

Sharpe asks, "What happens when we proceed as if we *know* this, anti-blackness, to be the ground on which we stand, the ground from which we attempt to speak, for instance, and 'I' or a 'we' who know, an 'I' or a 'we' who care?"[53] The weather, she argues, is a way to proceed in that knowledge, to talk about the total climate, and to understand how the totality of anti-Blackness conditions worldmaking from within the zone of non-being. As such, the weather for Sharpe is always "an indication of, and related to, the larger antiblack world" because it "registers and produces the conventions of antiblackness in the present and into the future."[54] Climatic expectancy is understood against the *certainty* of anti-Blackness, not strategic *enactments*, such that despite "predictably unpredictable" changes in the total climate, the weather will be one in which "Black people become the *carriers* of terror, terror's embodiment, and not the primary objects of terror's multiple enactments, the ground of terror's possibility globally."[55]

In effect, Flores thinks of climate more as a rhetorical ecology, or dynamic systems of addressivity that afford and sustain potential meanings and identities, where Sharpe thinks of climate more as an ecological rhetoric, or interdependent environments that attune inhabitants to the differentiated living and dying of racial terror, which includes and is expressed through established addressivities.[56] I find particular use in combining their visions of climate-as-expectancy to understand anti-Blackness as endemic to the worldmaking of Whiteness. Strategically, anti-Blackness is always available (it is in the air so to speak) and can be drawn on again and again, one intersectional variation after another, to reconfigure Whiteness antagonistically toward racialized, "affectable others." Ambiently, as a total climate, anti-Blackness is an inescapable condition that one inhabits because its force transcends its variations and functions as an attunement to Whiteness as a milieu. The strategic potential of anti-Blackness becomes ambient by being always already available, and ambience becomes potential strategy by attuning inhabitants to the conditions of an anti-Black world.[CI] Thinking climatically, Whiteness as a milieu can be explained both in terms of the weather that occurs and the relentless inclemency of the weather.

Climate as Trophic

Climate is about more than expectancy, however, and approached through biological and ecological science, climate directly shapes and is shaped by the cycles that make up webs of biotic and abiotic relations. Given prevailing conditions, some ways of being thrive and others do not, and those ways of being further impact prevailing conditions. In that regard, climates are trophic. As Diane Keeling and Jennifer Prairie explain, trophic dynamics are the complex relations by which different segments of an ecosystem nourish one another, often described through species behavior within food webs, although abiotic conditions that affect the systemic balance of nutrients are critical.[57] The inability of species to survive within climatic conditions has ecological consequences beyond whether they live or die, however. Trophism is about nested interdependencies—"the dynamics of matter and energy flowing through ecosystems" as Caroline Gottschalk Druschke

CI Even if I weren't suspicious of just about all instrumental thinking, I'm suspicious of "strategy" being quite the right way to look at the ambient nature of anti-Blackness, widespread though it is. Strategy has strong implications of "intentionality," which itself implies individual and volitional human action. And isn't ambience a dynamic and ephemeral composite, in some ways beyond individual control? How can one leverage ambience, mobilize it, get it to work for them? Aren't we always already working for (and in) it? But it does seem like Nate's getting at an important cultural logic here, maybe akin to what Val Plumwood calls the "logic of colonization," which may well operate strategically, but always does so through specific everyday techniques to maintain strong dualisms as normative. See Val Plumwood, *Feminism and the Mastery of Nature* (New York: Routledge, 1993), 41. In other words, there are specific techniques involved in cultivating ambient norms (certainly as much as the other way around), and these seem to operate "strategically" to the degree that they've become normalized as a cultural logic, which means the "strategy" is to continue their naturalization by not noticing it as strategy. I always learn from Nate, but I've also learned from Plumwood how to think about the ways ambient norms of anti-Blackness (anticolonized, antifeminine, anti-emotion, antinature, antihomosexual, etc.) share a dualistic logic of denied deference through specific techniques. Ambience, yes. Climates, yes. Ecologies, yes. But also the molar logics and molecular methods whereby they're created and maintained.

writes, of "things or forces or practices that robustly and tenuously hold together" a particular "ecological niche." Climatic conditions are trophic because they express and condition the "the myriad form of relations that hold us—a human and other-than-human and other-than-biotic 'us'—tenuously together."[58] Trophism is about relational worldmaking, niche by niche; however, the logic of nutrition that informs trophism deflects from the fact that trophism is inherently violent (what feeds on what and under what conditions), which raises the question of how forms of violence sustain and unravel ecologies.[JL] How do climatic conditions of violence affect worldmaking at the level of ecological provisioning, not just expectation?[BM]

Approaching climate through trophism moves the rhetoricity of climate more strongly toward extra-metaphoric, materialist thought, because strains of rhetoric would be understood in terms of their capacities and the conditions that support them, particularly violent conditions that debilitate "non-beings" to sustain those capacities.[59] Further, those capacities matter whether or not they produce some form of giving voice or being silenced; relevance is not limited to enabling particular entities to act through symbols.[60] Trophically, rhetorical capacities arise within violent relations that constrain and nourish what is and is not possible at an ecological level, which means that the focal point for anti-Blackness as a visceral attunement to Whiteness would be de-individuated to consider ambient dynamisms that sustain the milieu rather than to focus on experiential expectations and responses.

In *Ecologies of Harm*, Eatman broaches these issues, and although she does not discuss trophism *per se*, she argues that the "entanglement of violence and rhetoric produces and reproduces people, systems, and environments."[61] She

---

JL The question of what nourishes violence cracks open the local and asks us to look for sources upstream, in rainfall, embedded in underlying soil.

BM This emphasis on worldmaking raises questions about ethics too, which feels like a theme that threads into all of our chapters. If we think about the individual and, by extension, individualized senses of rights and responsibility as processes of territorialization (which I think you do earlier in this essay), how does this trouble the ethical frameworks and sensibilities shaped through these processes of individuation and territory? What does this mean for the ethical frameworks we've inherited? I struggle with this as I teach graduate and undergraduate students research methods and every time I go to put in an Institutional Review Board (IRB) application. What violence does the inheritance of institutionalized ethics do? What violence attends teaching and practicing research ethics otherwise?

elaborates on Sharpe's description of an anti-Black climate and its weather by exploring ways that "direct, structural, and cultural violence is hospitable to only certain identities and practices" in contexts of lynching, lethal injection, torture, and memorialization of racial violence.[62] She argues that public violence informs rhetorical ecologies and analyzes the "constant sensory presence" of violence, extrapolating beyond anti-Black violence.[63] For Eatman, rhetoric is broadly understood in terms of scenes, scripts, policies, enactments, spaces, practices, and so forth, and like Flores, she is deeply concerned with addressivity, particularly how forms of violence and rhetoric are productive of further such violence and rhetoric. However, Eatman stresses that "the context of violence . . . is also violence" and that navigating the constitutive effects of violence is a form of dwelling.[64] Dwelling aligns with trophism: Which ways of living and dying hold an ecology together or unwind it? Which ways of being are nourished and which are nourishment? She considers this question in terms of response-*ability*, or capacities to respond: "studying rhetoric from within violent rhetorical ecologies, rhetoricians must address which responses were available, to whom, and why, as well as how existing responses strain those boundaries."[65] In this way, Eatman extends Sharpe's work; for purposes of this chapter, you could say that anti-Blackness feeds Whiteness.[66] A trophic perspective such as Eatman's contrasts with studying the embodiment of rhetors (in)capable of normative symbolic competencies; focusing on Whitened and Blackened embodiments attends to performative identity, which is distinct from focusing on an anti-Black "horizon of death" that holds together relations necessary for a racialized milieu.

An important difference remains in that dwelling for Sharpe is about inhabiting a zone of non-being, which pertains not only to contesting the climatic limits established by ecologies of violence but also to the transformations wrested from and despite such a climate. For her, it is critically important to recognize that Black peoples are "constituted through and by continued vulnerability to overwhelming force" but "not *only* known to ourselves and to each other *by* that force."[67] The latter point is essential: violent relations are trophic, but what they tenuously hold together are not only those relations of violence. Eatman highlights the affordances and response-ability sustained by the violence of ecologies, where Sharpe stresses the resilient worldmaking of Black being that exceeds violence yet, nevertheless, is constituted by continued vulnerability to violence. "Here there is disaster and possibility."

Sharpe describes this worldmaking capacity by placing climate, weather, and ecology in "trans*formative" relation to one another: "The weather trans*forms Black being. But the shipped, the held, and those in the wake, also produce of the weather their own ecologies."[68] The asterisk in trans*formation signals the "range" of transformations "enacted on and by Black bodies" and "functions as the wildcard," inviting exploration of the "unthought" of Black life.[69] McKittrick might call these trans*formations Black geographies. In Sharpe's terms, "the weather is always ripe for Black death" yet "also produces Black resistances and refusals."[70] Trophically, the anti-Black climate of a White milieu is not accounted for by Black deaths and Blackened disablement but by attending to that inescapable, morbid ripeness that drives adaptations in ways of being including nourishing its refusals. As Tiffany Lethabo King observes, anti-Blackness "is not an event, not even a structure, but a milieu or active set of relations that we can push on, move around in, and redo from moment to moment."[71] As an example, Armond Towns analyzes the Underground Railroad as an environmental adaptation in response to Black death that resists and refuses Black death.[72] The violence does not eradicate the potential to be otherwise; it compels exploration of how to capacitate new forms of inhabitation without making trans*formative potential a bright-siding apology for or acquiescence to violence.[73] Sharpe thus provides nuance to the ambience of anti-Blackness: "When the only certainty is the weather that produces a pervasive climate of anti-Blackness, what must we know in order to move through these environments in which the push is always toward Black death?"[74] Ecological conditions affect the climate over time, and likewise, climate shapes ecologies. Mutual conditioning occurs across levels and places: climate to ecologies, ecologies to climate, but the experience is the weather—by minute, by month, by season.

Of course, knowledge of that "overwhelming force" looks very different depending on who "we" are in relation to the visceral attunement that anti-Blackness provides. Knowing ourselves for Sharpe is not knowing ourselves for me, and knowing each other not only by that force is more complicated still if one contrasts who is of the weather with who is doing the weathering. Eatman asks, "How can rhetoricians study rhetoric from within violent rhetorical ecologies?"[75] The answer is contingent on how the climate affects you, and "it is strikingly difficult to extricate ourselves from everyday practices this deeply built into our material infrastructure," which I speak to momentarily in the conclusion.[76]

Rounding out this discussion of trophism, an anti-Black climate characterizes the inhabitation of violent relations that help sustain Whiteness as a milieu. The question trophism raises, in contrast to expectancy, is how anti-Blackness as a climatic attunement feeds the worldmaking of Whiteness and the Blackened trans*formations of being that are driven by its weather. Such a perspective is not about particular embodiments and what they can and cannot do but about dynamisms and worldmaking as a generative set of conditions. The point of engagement is ecological rather than evental or agential: seeking changes in the relations that "hold things together" while understanding that the relations of violence do not only feed into or contest similar violence. Changes in ecologies means changes in trophism that fold back into the climate and its weather and, so, recursively change the climate.

Alright, Then What?

Where does this leave one grappling with Whiteness in its anti-Blackness and in terms of rhetoricity? Building on others' work, I have re-presented Whiteness-as-milieu as being about worldmaking, re-affirmed that it is anti-Black whatever else it may be, and added that expectancy and trophism provide complementary ways to think of anti-Blackness as a climatic attunement to Whitened worldmaking. One can expect anti-Black weather (violence, dispossession, nullification) of Whiteness. One can also see that an anti-Black climate nourishes and is nourished by the violent relations holding Whitened environments together, niche by niche.

What Nakayama and Krizek identified as strategic, I have approached as ecological. The milieu of Whiteness strategically secures the racial invention of humanness, world, and their relations, but expectancy construes invention in terms of anticipating form and performance at the level of possible intersectional enactments against the certainty, however variable, of anti-Blackness. Trophism understands invention through provisioning relations that are conditioned by but exceed violence. Crucially, Whiteness engages worldmaking differentially; no general inventive capacity is derived from it, only relational dynamics that are pervasively, but not exclusively, anti-Black. If accepted, there are many consequences to this synthesis of thought and many more questions, but two connected consequences I will end with pertain to incommensurable attunements vis-à-vis an anti-Black climate and the meta-ecology of Whiteness.

As to the first, Sharpe writes as one who must weather an anti-Black climate and knows the imperative of "seeing and imagining responses to terror in the varied and various ways that our Black lives are lived under occupation; ways that attest to the modalities of Black life lived in, as, under, and despite Black death."[77] Writing as one who is of the weather, the Whiteness I inhabit pushes toward Blackened death over and above my individual performances within the strategic contours of White dominance. As a case in point, I can quote evocative language and arrange abstractions to sound minimally conscious of violence I will not suffer but in sympathy. I can write in awareness of privileges and protections that accompany my cisgendered Whiteness. I can gesture to a range of anti-Black weather: from Black artists' coerced into White visions of Blackness to see their work produced;[78] to Black intellectuals seeing their contributions ignored or particularized to reflect only on Black experience;[79] to Blackened communities uprooted to make space for infrastructure or suffering undrinkable water;[80] to Black bodies abused under medical authority;[81] to Black presidents turned oozing zombie targets for gun "enthusiasts";[82] to Black life made invisible by hyperbolic presence as entertainment;[83] to Blackened suffering seen as generic, not individual;[84] and, agonizingly, to Black folk murdered again and again under color of law—in their cars, in their beds—followed by gratuitous rehearsals of the debate, "Is *this* killing racist?"[85]

But there is a texture of being decidedly beyond what I can touch because I do not inhabit a zone of non-being; rather, I help drive the climate and its weather because I inhabit a zone of (White) being and so my attunement to an anti-Black climate is thus incommensurable with one who must weather it, even if I may be harmed by it in different ways.[86] Keeping in mind that Whiteness, and indeed raciality, is uncontained by anyone's corporeality or subject position (let alone mine), my participation in anti-Blackness is not for me to grasp fully and certainly not mine to self-determine; as a participant, my relationality does not end where another's violent act begins because Whiteness is more than a series of acts. One can't be of the elements and command them, so my racial attunement via anti-Blackness (both its potential to persist and to be otherwise) arises from my affective immersion in the climatic conditions of Whiteness.

Said differently, I cannot account, as an agent, for the conditions that offer being to me. The agencies I embody to be a scholar and teacher of rhetoric, even one who yearns to be antiracist, depend on and activate the expectancies and trophic dynamics of a Whitened milieu, so anti-Blackness differentially conditions the ambient rhetoricity of dwelling in the study of rhetoric, beyond directly addressing race or one's racial identity. I not only inhabit a different zone of being from Sharpe; I also dwell in relation to her, to Black life and anti-Blackness, through forms of violence that sustain that very difference even in resistance. As part of a Whitened milieu, the study of rhetoric is ecologically anti-Black and that raises questions that surpass what privilege, complicity, and other agentic foci can explain when trying to change Whitened worldmaking through scholarship and education about rhetoric. Hence, although I speak only of my own limits, one who inhabits a Whitened milieu of necessity dwells differentially in an anti-Black climate, conditioned by its expectancies and trophism, which is not about a racial binary but about meta-ecological conditions.

Reflecting on Sharpe's insights in terms of my chosen field and what my own limits can disclose, I believe Whiteness operates climatically as an affective mode of modes, a kind of meta-ecology wherein characteristic, pervasive anti-Black brutality and disablement provisions the formation of a multiplicity of ecologies, across which strains of rhetoric arise that are part of the persistence and collapse of those ecologies. That ecological multiplicity includes new niches that Sharpe argues Black people establish in refusal of anti-Black violence. She asks: "In the midst of so much death and the fact of Black life as proximate to death, how do we attend to physical, social, and figurative death and also to the largeness that is Black life, Black life insisted from death?"[87] It also includes the layered, Blackened conditions of deportability and disposability of Mexican American bodies that Flores discusses. It includes complex ecologies of harm where anti-Black violence circulates in excess of the targeted suppression of Black life, which Eatman discusses. It includes the niches where academic institutions are embedded (in communities, in places) through relations of knowledge production. And still more because anti-Black violence enables relationalities that are not only of and about anti-Black violence but are also vital to sustaining Whitened worldmaking in its various forms—and, also, to transforming it. Thinking climatically is meta-ecological, such that Blackened bodies and Blackened geographies sustain and unravel Whiteness within and across different niches over and above the identities of people inhabiting different ecologies. In that, I find McKittrick's language of trajectories useful for capturing the decentralized meta-ecology of pervasive anti-Blackness and responses to it.

However, in parting I want to throw open the ramifications of a climatic understanding of Whiteness because ecology and climate should not be seen as figures of exclusively cultural institutions and practices; rhetoric does not exist on a plane of culture separated from other material conditions. The issue of climate-as-metaphor has lurked on the edges of this chapter, but I name it now to step out from the arguments presented here and to interrupt any sense of containment one might feel in adopting a climatic view of Whiteness in its anti-Blackness. Seeking to transform an anti-Black climate, Sharpe insists, goes beyond forming direct responses to the violence because it involves unsettling deep relations of racialized ableism across scales (indeed, forming those scales)–climate to ecologies and ecologies to climate, which engages many dimensions of worldmaking. Adopting Gottschalk Druschke's point of view, unsettling such relations does not occur apart from "the myriad form of relations that hold us—a human and other-than-human and other-than-biotic 'us'—tenuously together."[88] In that regard, the work of unsettling Whiteness as a milieu "must go beyond human doing" and the (dis)abilities of humanity and its others because a milieu is not reducible to agentic control over symbolic resources; as Rickert states, it is not possible to "simply and directly *choose* to dwell otherwise."[89]

Paying heed to anti-Blackness as climate enables understanding of anti-Black raciality and intersectionality as a meta-ecological dimension of a Whitened milieu, but it also leads one irrevocably to confronting how ecological relations between humans and other-than-humans are enlisted in the anti-Black worldmaking of Whiteness and the ways anti-Blackness directly and indirectly impacts the livability of all things. Anti-Blackness permeates resource extraction, public health, property relations, waste disposal, technology production, urban planning, etc., etc.; and, as weather, it continually impacts all kinds of beings in and beyond Blackened communities, in and beyond people who think that they are White. As Kathryn Yusoff puts it, "When the storm is over, there will be another. The storms are always coming, with faster and greater intensity."[90] Transforming the worldmaking rhetoric of Whiteness means confronting the daunting range of relations that constitute a world engulfed by and for universalized, self-overcoming humanness, and that involves understanding how anti-Blackness threads through that world—in the expectancies of its inclement weather and the trophic relations that provision it. The oblivion Baldwin warned of was not figurative, after all, so forming new ecologies means forming new ways of inhabiting relations with otherness in the fullest sense.[CR] In the next chapter, Jennifer explores ableism in ways that further complicate Whiteness as inclement weather by directing our attention to the motivational affect thrumming in stories of disabled people surmounting their bodies: benevolent ableism is what a climate of normative legitimacy for Whitened ableism, laundered through inspiration, feels like.

CR The sheer labor required to even slightly wobble the "daunting range of relations" that sustain a racist ecology is overwhelming and unimaginable—especially with full awareness that any shift simply means recontouring trophic relations, doing nothing to eradicate violence once and for all, but only to shift the forms of violence or who or what carries the weight of it as expectancy, who or what is nourished or is nourishment. In the context of institutional transformation that I speak to later, I have some touchstone ideas and figures to inspire ethical orientation and praxes in the face of this vastness, but I wonder, friend, what do you lean on to help you face the oblivion?

Microclimate 2

(Nate and Jen)

I have been watching shows about ancient cities and civilizations hosted by an engineer who has a prosthetic foot. Albert Lin brings advanced imaging technology to bear on the process of unearthing secrets. His foot, an engineering marvel itself, is featured frequently in ground shots as he enacts a latter-day Nat Geo version of Indiana Jones. I am left to wonder what his metal foot means for the discovery of ancient secrets these shows offer. I like Albert and my reaction no doubt demonstrates why he is the host, but I really do like him. However, the camera keeps making his foot a viewpoint for experiencing archeological discovery, which is not about doing archeology per se but about overcoming obstacles. The unsaid suggestion is that the imaging technology that allows one to see what one otherwise could not also is about overcoming. His foot adds a layer of unexplained inspiration that has no direct pertinence to learning about Nan Madol or Mayan temples, but which says something anyway. Lost things can be retrieved via technology.

The visualization of Lin's foot, riding the line between genuine vulnerability and projected pathe-ticness, demonstrates how ableism is the legitimating credential of Whitened virtue. The world's ability to be made and remade validates its making and remaking. Yes, coloniality was a cavalcade of crimes and horrors, but look at what "we" can do. Look at "us" retrieve what was destroyed or broken so that "we" may revere it. The lie of such a narrative is that the process of fixing or reparation will ever reach a point of recognizable completion. Anne Anlin Cheng sees the failure of "liberal racial rhetoric" as its inability to "tolerate the possibilities of subjective failures or corporeal ambiguity on the part of its recuperated subjects." Or, in a world of Whiteness and ableism, only the center is allowed to continue exploring and evolving, whereas any instability of selfhood at the margins muddles the triumphant endpoint for the assumed "we."

It seems unavoidable to us that Whitened worldmaking would suppose abled worldmaking, then, because both carry a sense of rightful, necessary, and better order that is legitimated by the supposed transcendence of the pain and struggle of overcoming. The resulting climate is one in which attention is spent unevenly, glancing off any surface that provides more than the briefest resistance to feelings of rightfulness, necessity, and order. But it is more than Whiteness and ableism presupposing each other; each offers balance to the other. The former emphasizes control over the self and the wider world while the latter envisions unfettered potential in ideal embodiments. The bleaching effect of Whiteness attains a sense of benevolence through ableism that moralizes and naturalizes continued resource extraction and its cruelties. For instance, in Flint, Michigan, the lack of care/attention to the future of these Black bodies' lead-lined capillaries and brain matter, their altered neurological futurities, is not just about socioeconomic inequities but also the assumption that debilitation is inherent in those bodies over there. The kicker is that, in this worldview, outside aid is no longer rightful reparations but altruistic philanthropy.

Benevolent ableism shapes expectations for disability so that people with disabilities are a reminder to ableds to feel like they can do more or should be more thankful for what they have. And the omnipresent narrative of disabled overcoming that inspires ableism provisions the climate of pathology that establishes values incumbent to ableism as a milieu of benevolence. There is a sense of obligation embedded in this benevolence, as if disabled people and disability generally surround and oblige ableds to do something with their self-gifted sense of good fortune—inspiration as moral detergent for feeling uplifted by one's assumed superiority.

So, a formula for inspiration: extract power from limitations that are unrecognizable except through the critical pathos of pathology and then watch as that power enables those limitations to be transcended. From inability, constraint, loss, and pain comes ability, freedom, surplus, and joy. Of course, the formula needs raw material to produce value, and that is the function of the people who are pathologized. They feed a vision of the normal that is actually a state of pathology transcended; such lives trophically support a climate of benevolent (redeemed?) ableism. The consumption of disabled inspiration porn feeds the climate that expects it. Critiquing this pathology repurposes the pathological, however, so there is no simple rejection on hand. Rather, such critique requires the same sort of tenacity that living in a state of extraction requires, where the project of living is one of enduring anticipation.

3 Climates of Benevolent Ableism

(Jennifer)

In a Ted Talk in 2014, the late disability activist and writer Stella Young voiced what might be one of the best summations of the problems with the current climate surrounding disability. In her talk, "I'm Not Your Inspiration, Thank You Very Much," Young recounts being interrupted at a lecture by a student who wanted to know when she was going to get to the real point of her speech; "You know, like, your motivational speaking. You know, when people in wheelchairs come to school, they usually say, like, inspirational stuff?"[1] She uses this anecdote to illustrate how the dominant portrayal of people with disabilities, "inspiration porn," reduces the experience of disability to an emotional thrill for the presumably abled viewer. In inspiration porn, short images and/or video clips show individuals as "overcoming" their disability temporarily, often to achieve some sort of athletic triumph. A girl with no hands learns to draw by holding a pencil in her mouth. A small boy with prosthetic legs beams beatifically as he runs around a racetrack. These images, although disparate in origin and distribution, have saturated societal understandings of disability to the point where these students viewed Young not as an adult individual in her own right but as an inspirational figure in narrative meant to uplift others.[CI]

As narrative "is an important form through which lay populations come to understand disability,"[2] the images, video clips, and articles that coalesce under the umbrella of inspiration porn are key to sustaining expectations for how disabled

CI Reader, can I just pull you away for a second to remark publicly (if the notes in a book qualify) about how marvelous it is to dwell in Jennifer's prose? And watch the move that she makes next: to situate the importance of narrative for understanding disability. Just don't think for a second that this whole chapter isn't also enacting the importance of narrative for the integrity of scholarship at large. Across cultures and traditions, stories are a way to make sense of the world, your own and otherwise. Jennifer knows it and shows it.

bodyminds are supposed to act in the public sphere, primarily as examples that inspire "able" individuals to work harder and achieve more. Inspiration porn centers on representations of disability that show "impairment as a visually or symbolically distinct biophysical deficit in one person, a deficit that can and must be overcome through the display of physical prowess."[3] Within the larger ecology of "spreadable media ... discourses of disability associated with charity, inspiration and existential guilt" are typically packaged as epideictic discourse, praising the portrayed individuals for how they deal with their disabilities.[4] Yet this praise is contingent upon a substitution of heartwarming affect for sociorhetorical agency. The genre's misconceived portrayal of disability is summed up in Young's statement, "No amount of smiling at a flight of stairs has ever made it turn into a ramp. Smiling at a television screen isn't going to make closed captions appear for people who are deaf. No amount of standing in the middle of a bookshop and radiating a positive attitude is going to turn all those books into braille."[5] Inspiration porn's ubiquity rhetorically orients our attention toward disability as an ontological state that can be made to disappear, or at least fade to the background, via praiseworthy physical labor and positive affect.[BM]

The focus on these stories of disabled individuals and their positive attitude is made possible within and contributes to a climate of benevolent ableism, an affective orientation toward embodiment that only recognizes disability when it is yoked to narratives of emotional uplift.[NS] Ableism is typically defined as a broad "system of discrimination that rhetorically invents and employs the idea of a 'normal body' and treats physical deviance from that norm as lacking something that all other nondisabled people share."[6] Within this societal structure,

---

BM The emphasis on inspiration makes me think about how the root word, inspire, means breath or breathing. I take my own ability to breathe for granted and breathing becomes a background condition that I don't pay much attention to most of the time. Your argument in this chapter, and especially what you are saying about how inspiration porn and the ubiquity of benevolent ableism render disability invisible, helps me remember the contingency of breath in such climates. Breath is easier to forget for those who benefit from climates of ableism and whiteness.

NS There is so much richness to this that you are touching on. The idea of benevolent ableism makes me wonder to what extent disablement and debility are intrinsic to persuasion. If motivation is connected to uplifting, overcoming, fixing what is broken, I would like to understand how the history of ideas of persuasion is also yoked to ableist thought and practice, in the sense that inspiration porn is a refined genre that emphasizes what is baked into ideas of motivation generally.

disability is framed as a "pathological variation, as deficit, and, significantly, as an individual burden and personal tragedy."[7] While many analyses of ableist systems and structures focus on the overtly harmful nature of these, such as the labyrinthian processes of securing academic accommodations,[8] these hostile orientations toward disability are situated within a broader climate of benevolent ableism, in which "common sense" dictates viewing disability as something to be overcome and disabled people as those who need help from ableds in finding this path. For example, activist Imani Barbarin tweeted "I think about the time an abled random stranger threw my crutch into the pool 'to help me swim' a lot."[9] The thousands of replies (marked with #AbledsAreWeird) told similar stories of ableds performing illogical or sometimes even harmful actions that were aimed at prompting the person with disabilities to do more, vaguely defined, and overcome their use of necessary aids like crutches, wheelchairs, and canes. While it would ordinarily be considered ludicrous to throw another person's belongings into a pool unprompted, what gives these otherwise incoherent actions structure are the attitudes toward disability embedded within the climate of benevolent ableism; the goal should always be to help surmount disability's impact on the body, rather than alter the range of structures or beliefs that frame said body's existence as deficient.[CR]

The example of #AbledsAreWeird encapsulates how a rhetorical climate[CI] "shapes a certain logic of belief and subsequent action" through "feelings,

CR Your conversation here so starkly highlights the violence of "well meaning" ableist narratives of overcoming and benevolence. Not only do these narratives work to invisiblize their harm behind good doing, they normalize and invest in a fictitious narrative of the possibility of overcoming disability that faults individuals for "failing" to "rise above," rather than invest in transforming these exclusionary structures and ideologies. Your beautiful writing and lucid analysis help me see and feel how this happens.

CI Well after we'd long all gone running with the "rhetorical climates" idea, Jennifer casually mentioned to me in an email that, all this time, she'd been understanding our project through the lens of Ellen Gorsevski's work on climates-as-textual, and is that what I had in mind? Wait, what? I had no idea someone else studying rhetoric had ever written about climates as rhetorical texts, at least not so explicitly. Jennifer deserves a big high five for that. And Gorsevski, of course, deserves enormous credit for developing the notion in the first place. While I don't think it would be accurate to say Gorsevski's work on the topic had a conscious influence on *Rhetorical Climatology*'s ideation for most of us—we never read it in our reading group anyway—it's an important early move in this direction and probably deserves more than a hidden note. Thank you, Dr. Gorsevski. And thanks, Jennifer, for citing her up ahead.

intuitions, and experiences."[10] In the wake of #BlackLivesMatter, it is easier to see the multiple ways in which our racial climate is hostile, in which "the weather is the totality of our environments; the weather is the total climate; and that climate is anti-black."[11] Yet a climate that is anti-Black is not separate from a climate that is antidisability. Rather, the pathologizing of bodily features as medical defects is but another version of "the judgment and classification that subjected young Black women to surveillance, arrest, punishment, and confinement" under the guise of correcting "wayward lives" and bodies post-Emancipation.[12] Considering the rhetorical impact of inspiration porn is to more deeply consider how the underpinnings of multiple forms of marginalization are indebted to unexamined warrants about which bodies are more "defective" and therefore more exigent.

To live "in the wake" is to live a life that is not allowed to be understood as is.[13] These fundamental misunderstandings beget misrepresentations, and vice versa.[CR] Genres are coalescences of climate stuff, the dominant affective attitudes and cultural that reside in and motivate everyday genres. Within genres, "individual intentions and socially objectified exigences mutually produce and sustain each other," which means we can look to mainstream genres to better understand what configurations of bodies are understood as most socially profitable.[14] Or, the rhetorical power of genres is their encapsulation of dominant climates that further legitimate such atmospheres when used and circulated. The genre of inspiration porn supports a future-focused, ableist climate in which the disabled body has worked hard enough to transcend the category of disability, setting up at least partial erasure of disability as the presumed happy ending. To analyze how inspiration porn suffuses the climate with benevolent ableism, I start by contextualizing the contingent status of disabled individuals as rhetorical actors. I then re-examine the generic features of inspiration porn in order to excavate the latent assumptions and fears about disability and embodiment. Lastly, I parse how multiple instances of contemporary ableism and their overt

---

CR This is so it—to ask how a particular racist, ableist, and otherwise toxic climate is creating conditions that do not even allow a life to be understood, seen, registered just as one is, with no available receptors to even sense a person as a human, as fully alive. If your body or lifeways deviate, it is to be invisible, to not exist, but through the existing narratives and frameworks yoked to power, dominant ways, if not also to be eliminated, violated, worked out of existence in one way or next. Or be punished for trying to survive or exist on one's own terms, in one's own body.

points of bias are linked to the seemingly kinder, gentler portrayals of disability in inspiration porn. The accumulations of these patterned forms of action and response show how, despite claims of increased inclusion, the current climate is one that still automatically assumes deviations from the bodily norm need to be compensated for with ableds cheering from the sidelines.[JA]

The "Newness" of Disability

The formation of U.S. immigration policy was originally grounded in overtly discriminatory "notions about what it means to be fully human [that] merge race and disability as categories of deficiency."[15] The resulting legislation attempted to keep those with "feeblemindness," assumed to be more prevalent in particular ethnic groups, from sullying the new country's ostensible purity.[16] Even in more contemporary media, the "deviance is evil" trope seen in movies like *The Exorcist* magnifies repulsion for physically deviant bodies through linking them to non-Western locales and occult practices.[17] The beliefs that pose physical disabilities as mirroring some form of inner immorality or nonhumanness are inextricable from metrics of racial and ethnic exclusion. Of course, the original fears of contaminating the nascent U.S. nation with disability still echo in contemporary panics around embodiment and disability. Mel Chen's work on Chinese manufacturing of children's toys and the potential presence of lead contamination during the 2000s shows how arguments about the risk of disability to the individual, specifically IQ loss, were scaled up to the level

JA Strike this comment, if you see it off center from the crystal clarity of your argument, Jen, but the disabled "otherwise" world is alive, just as is the "otherwise" world of Black life mentioned earlier: antidisability, anti-Blackness cut from the same white bolt. I simply am reminded of the spike of disability-identified students on my campus (17 percent of my enrolled students are such), consistently because of hyperanxiety, attention deficits, performance anxiety, loneliness, and so forth, during times of mass hysterias and lies, rising tides, and authoritarian zeal. I do not mean to enact the privileged move to "innocence" ridiculed by Eve Tuck and K. Wayne Yang in "Decolonization Is Not a Metaphor," *Decolonization: Indigeneity, Education & Society* 1, no. 1 (2012): 1–40. One would not feel more Indigenous or Black or disabled through well-intended compassion and exposure. Yet my students, like yours, are struggling, however they are labeled by campus. Which is why I find responsible hope in the concluding comment—still ahead, readers!—from Kuppers and you in this chapter, to "find comfort in the company of others whose pain might be different, but who somehow feel sympatico" while respecting "incommensurability." The work is long and long overdue.

of the nation-state, resulting in discourses of racial contagion that threatened the privileged (White) American body with diminished ability.[18] The disabled body operates, as Nate puts it in the previous chapter, trophically, feeding racial fears of Chinese manufacturing via hyperbolized narratives of generational disability.[CR]

Even without overt ethnomoralization, disability has been used as a rhetorical device to identify certain bodies as only legitimate in marginalized, clearly cordoned off spaces. Throughout the nineteenth and well into the twentieth century, those deemed as having insanity, "the catch-all term for mental, social, and cognitive differences," were regularly institutionalized in asylums on the authority of medical professionals, often at the behest of family members, in order to keep the rest of the populace safe from this contamination.[19] When disability was allowed into the public sphere, it was to shore up bodily hierarchies. The nineteenth-century freak show "dramatized the era's physical and social hierarchy by spotlighting bodily stigmata that could be choreographed as an absolute contrast to 'normal' American embodiment."[20] The history of disability is a history of confinement and exploitation, determined by abled desires.

It is as late as the twenty-first century that understandings of disability have started to move away from these overt forms of othering, although, as with many cases of social change, the seemingly more positive depictions nonetheless still bear cultural anxieties about potential upset to the social order.[21] The positive shifts largely center on increased representation; disability theorists write *New York Times* columns about disability-related issues (Garland-Thomson), popular TV shows have main characters who use wheelchairs (*Glee*, *Superstore*) or are autistic (*Parenthood*, *Sesame Street*), and the term "ableism" is often present on lists of social ills alongside the better known "racism" and "sexism."[22] The expectation is that schools, stores,

---

CR Trophism is a thread throughout our collection. It is both frightening and enlightening to see it come up here in this way. In tracing who or what is nourished or nourishment (or put another way, whose well-being is supported, centered, and blooming and whose lifeways and life is diminished, harmed, or obliterated) within a climate does not evoke narratives about overcoming or eradicating violence. I've been thinking about trophism as part of a critical praxes and ethical orientation that aims to name/understand the conditions of survival and shift them with intentionality, accountability, and in view of one's complicity and the constancy of violence. This is a strong throughline I notice in our collection and conversations. It is hard to sit with sometimes.

and legal structures will accommodate those with mental and physical disabilities. However, this also means that in a relatively short amount of time (the Americans with Disabilities Act has only been in existence since 1990), there has been a seismic shift in the cultural practices related to disability, a shift from the belief that certain bodies are inherently dangerous and must be removed from public spaces to a social attitude that diversity includes celebrating physical and mental difference.

These societal struggles are not just over competing hierarchies of disability and race but over epistemologies of the self. The "new normal" of disability is that "more people are living with disabilities as a result of medical technology and that we perceive more conditions as disability as a result of the normalization of the category."[23] On the one hand, the increased social awareness of the ubiquity of disability works to destabilize some ableist norms. However, this increased awareness and presence of people with disabilities in the public sphere also undermines the long-standing assumption that ableds are defined through their distance from disability. The disabled figure has typically been "the vividly embodied, stigmatized other whose social role is to symbolically free the privileged idealized figure of the American self from the vagaries and vulnerabilities of embodiment."[24] Now, the increased presence of disability in everyday life threatens to rupture normal understandings of the relationship between bodies, their physical capacities, and social hierarchies. In the new normal where disabled people are more present and represented as equal to ableds, inspiration porn shoehorns disability into familiar bootstrap narratives, shifting attention away from the disability itself and toward the respectable striving of the individual.

Like most underdog narratives, inspiration porn leans into a generic familiarity where "characteristic tropes, devices, icons, or storylines" promise "an anticipated affective experience" of triumph, which in this case necessitates the reduction of disability.[25] The "typification" of these affective truths within common genres reinscribes tacit societal priorities, as seen in what is deemed in need of fixing; the "problem, as disability studies has persuasively demonstrated, lies in *how* variation and difference is construed as abnormal and 'wrong,' as a personal deficiency needing to be 'fixed.'"[26] The ubiquity of this genre indicates how fears about the limits of the self and uncertainty about bodily hierarchies have triggered a cultural need to self-soothe in the form of a recurring, nameable genre.

Inspiration porn's reliance on positive affect masks how its success "depend[s] on our ableist culture's low standards for the lives of disabled people";[27] the viewer is assumed to be operating from a non-disabled perspective but nonetheless holds a benevolent willingness to vicariously experience an ersatz version of disability. For example, inspiration porn often relies on the "supercrip" narratives of "disabled people who are represented as inspiring or extraordinary for performing both mundane and exceptional activities,"[28] usually in an athletic or otherwise physically demanding setting. The dominance of these narratives limits the range of social roles for disabled individuals and reproduces an understanding of bodily empathy as conditional upon meeting a certain bar of hyperachievement. The similarity with other spreadable media genres—such as underdog stories of eventual financial success, the "homeless to Hollywood" tales that go viral—means that one of the most accessible narrations of disability focuses on the self-reliant subject cultivating virtues of physical "grit" toward overcoming. The fact that these with "deviant" bodies and minds can see fit to smile in the midst of their unspecified suffering is meant to exemplify this warrant taken to its logical conclusion; the further you are from the heroic finish line, the more of a triumph you must experience when you get there.

Athleticism as Cure

Rather than address the medical, financial, or structural aspects that make it difficult to have a certain embodiment in particular situations, the narratives within inspiration porn rapidly move to a resolution where the issues surrounding disability are cured by individuals in a particular situation doing the right thing, often with "non-disabled people 'helping' disabled people."[29] Similar to other "feel-good" genres that rely on bootstrap narratives and focus on individual action as the cure to systemic issues, the viewer knows at the beginning of the video that there will be an emotional resolution, and this promise of resolution effectively removes the potential discomfort of encountering disability. Eli Clare lists some of the memetic forms of the productively laboring disabled body that overcomes adverse circumstances; "A boy without hands bats .486 on his Little League team. A blind man hikes the Appalachian Trail from end to end. An adolescent girl with Down's syndrome learns to drive and has a boyfriend. A guy with one leg runs across Canada."[30] While the discussed actions might be inspiring on their own, Clare skewers the tendency to narratively render the action and the disability as one and the same, a rhetorical move that supplants individual personalities with a medical diagnosis.

Inspiration porn pivots on the presentation of novel physicality to a presumably non-disabled audience. Garland-Thomson discusses how the freak show rhetorically functioned as a reinforcement of American exceptionalism as played out in the singular body; these shows set up a bodily binary where "the American *produces and acts*, but the onstage freak is idle and passive. The American *looks and names*, but the freak is looked at and named."[31] In contemporary examples of inspiration porn, we see a slightly altered version of this equation where the disabled body is still largely without agency to name itself, but this lack is presented as compensated for by the implied physical "production and action" that must have gone into the now inspiring result. The disabled individual is too busy running races, marrying attractive people, and lifting weights to have time to inform the viewer of what further desires they might have. The continued reinforcement of disability equaling hyper-achievement presents the only possible social relationship between disabled and able-bodied individuals as one of benevolent spectatorship. Just as "respectability politics" demand exceptional performances from Black individuals as a condition for acceptance, inspiration porn demands that disabled people occupy a body that strains toward a mainstream identity in order to belong.[NS]

The future-forward climate of benevolent ableism is supported through inspiration porn's tendency to focus on young individuals, often children or teens, as potentially less-disabled adults. These stories are framed so that the individual is seen primarily through the lens of their youth and disability, as in the headlines "4th grader with Spina Bifida almost couldn't attend a class trip. A teacher carried her so she could" and "Teen couple with Down's syndrome voted prom king and queen."[32] A highly disseminated inspiration porn meme of a young boy, perhaps four or five years old, exemplifies the issues with this sort of framing. The boy is captured mid step, his back prosthetic foot providing

NS Reading this, I am struck by the possibility that prosthetic logic is reversed with hyperathleticism. The "freakish" ability demanded is a prosthesis-of-excess that may or may not be attached to a technological/pharmacological prosthesis, shifting the sense of what supports what—the inspiring fanaticism of the person running too many races is the support that abilizes someone, lifting them beyond any disability. Any technological/pharmacological prosthesis is in fact what is supported by a hidden hyperability of spirit that overcomes matter.

leverage as his front one twists with the momentum of his stride. He is on a running track with a race number pinned to his chest. The caption that usually accompanies this image is a variation on "Your excuse is invalid," or alternatively, "The only disability in life is a bad attitude," a quotation attributed to Olympic ice skater Scott Hamilton.

Even setting aside the use of the term "invalid" as a descriptor for a child with disabilities, the athletic triumph is clearly set up as a form of compensation for the boy's disability. Despite being in the midst of a very difficult physical endeavor, the positive emotion of the boy's smile and the joy of physical competition are yoked together into a one-dimensional understanding of physical success as evidence of a "good" performance of disability. The use of a young child, innocent, with blonde hair and teeth that have not yet seen braces, also implies a temporality in which disability falls away as the child ages; if he is performing this well at age four or five, what will he be doing/how will his body be different at age twenty? Conversely, the blame side of this epideictic equation is made clear via the caption; bodies that choose to not maximize their physical potential fail and become "invalid." The only "real" disability is a lack of trying.

These positive portrayals of disability as an abstract fantasy spatially divides society in a way that "replicates the social untouchability of disabled people" and that affirms the idea of "the to-be-looked-at rather than the to-be-embraced."[33] The individual is not bragging about how they managed to do such amazing feats as run a lap with prosthetic legs, being named prom royalty, or even getting a conventionally attractive person to date or marry them. Rather, a benevolent outside observer has captured a "candid" moment of anthropological import, praising the individual body that is frozen in time at a high-performance moment. Just as most people do not typically meet Olympic medal winners or hand out with NFL stars, disability is to be placed on a pedestal and wistfully admired from the stands. The intellectual or emotional journey of the portrayed individual is reduced to a one-dimensional emphasis on the joy of winning that is meant to be contagious to the viewer, pulling on existing knowledge of other athletic triumph–related genres and that affective memory. The possibilities of looking at disability narrow to searching for supercrip greatness.

This snapshot of physical triumph also abstracts disability away from bodies and into a binary framework of moral failure versus redemption, reinforcing the idea that the disabled body at rest is performing itself wrongly. In this view, disabled individuals can occupy different levels of virtue as determined by their level of will to overcome, and therefore, failing to push one's body to the ultimate limit is an improper way of being disabled. If even the disabled are able to overcome inherent bodily weaknesses, then ableds have less to fear from their own vessels. Or, the able audience member has evidence for "overcoming the adversity of impairment, eating a perfect hamburger, or experiencing a well-functioning airport" as all equally fantastical daydreams.[34]

As Ahmed points out, some "signs, that is, increase in affective value as an effect of the movement between signs: the more they circulate, the more affective they become, and the more they appear to 'contain' affect."[35] The continued recirculation of these representations of disability do more than mask issues of inequity and access. Rather, the centrality of the positive affect of inspiration porn and the embedded assumptions about what it means to ethically inhabit a body supports a climate in which certain bodies are deemed acceptable and others are deemed disappointments. Beyond these memetic images of disability, this binary of awe/disappointment has also crept into broader political discourse and is used in support of political and social disidentification.

Climatic Effects

Many scholars of political rhetoric have made striking arguments about the rhetorical underpinnings of the intensifying political fractures of the twenty-first century.[36] What has been less discussed is how the language of disability has been weaponized as part of this fracturing, used to not only insult one's opponent but also to diminish their ethos through the specter of disability. When looking at examples in aggregate, what becomes clear is how the same assumptions about what counts as normal embodiment motivate the seemingly more positive affects that are central to inspiration porn. In these cases of political warfare, one's sociopolitical opponents are assumed to have not done the work of overcoming their presumed disabilities, and as such they become legitimate targets for ire. The following examples demonstrate how the climate of ableism that legitimates accusations of disability as part of everyday political discourse operates from the same rationales that underpin the affective appeal of inspiration porn.

While ableist language is nothing new, the political fracturing surrounding the election of former President Trump in 2016 and successive widening of the gap between the American right and left was accompanied by an uptick of overt examples of equating political enemies with disability, particularly mental illness, as a valid rhetorical move. Trump is well known for feeding the climate of ableism, openly mocking the reporter Serge Kovaleski's physical disabilities with crude pantomimes. In the first presidential debate of 2020, many were angered by how Trump appeared to be weaponizing the fast pace of the debate to reveal Biden's stutter, following tweets where he made fun of "Sleepy Joe's" appearance and questioned if he was on drugs. Many prominent Trump supporters also used ableist language to attack figures on the left, as when conservative pundit Michael Knowles called activist Greta Thunberg a "mentally ill Swedish child."[37]

However, the left is not blameless in this weaponizing of ableist language. In early September 2020, the hashtags #TrumpIsNotWell and #TRUMPStroke went semi-viral as armchair sleuths squinted at videos, looking for signs of weakness or drooping that might indicate a series of rumored ministrokes. Countless insults were coined, making fun of Trump's appearance, particularly his weight, in a way that directly linked these physical attributes to his mental capacity and moral state. There have been numerous think pieces speculating as to the state of Trump's mental health; many going so far as to offer possible diagnoses, placing whatever the author finds disagreeable about his behavior in a medical frame with a definitive label. Following the storming of the U.S. Capitol on January 6, 2021, several psychologists called for the removal of Trump from office, arguing that he was inciting "narcissistic symbiosis and shared psychosis" in his followers.[38] Although one might argue that the president's diminished mental capacity is worth discussing, the rush to leverage a medical diagnosis in conjunction with moral deviance as a means of attacking the opponent's ethos reveals how disability is still viewed as an insurmountable deficiency. If one follows such logic to its endpoint, the use of a psychiatric diagnosis to attack Trump only works by invalidating the agency and will of his followers, delegitimating their humanity via accusations of disability. To an extent, the motivations of Trump or his opponents in using such attacks is beside the point. It is the continued utility of ableism as a rhetorical tool that demonstrates how the current climate is still saturated with the equating of disability and ethical deviance.[BM]

BM Your critical analysis here reminds me of what I've seen you do elsewhere, like in your analysis of the Pussyhat Project and the multiple works in which you address raciality, Asian embodiment, and normative discourses about food. I'm in awe of your ability to push back the curtain on normative discourses that are so ubiquitous as to be made invisible or that go against the grain of, in this case, a liberal political identification. Readers, see Jennifer

Beyond attacking prominent political figures, this weaponizing of ableist language has been employed to foster disidentification in ways that echo the crudest of biological inherency arguments. In the early months of the COVID-19 pandemic, many citizens of Wisconsin protested the restrictions by going to the state capital to try and pressure the governor to open schools and businesses. While the protestors were criticized for ignoring social distancing protocols, the critiques that gained traction very quickly took on an ableist tone. Tweets questioning the protestor's "mental capacity" and the hashtag #COVIDIOTS were retweeted and incorporated into more traditional news media coverage. People's fear and anger about coronavirus quickly manifested in the repeated use of terms like "stupid," "nuts," "morons," and "mental." In response, disability studies scholar Sami Schalk wrote an excellent breakdown on Twitter why such language is not merely offensive but part of a long legacy of using ableist language as rhetorical props to isolate, exclude, and incarcerate those deemed less worthy. In her words, the use of these labels to mark one's political opponent feeds into a long tradition of suggesting that "people with psych [*sic*] disabilities are inherently dangerous, disposable, dismissible, etc." and "relies on the oppression of [those with psychological disabilities] to be effective."[39] More than mere insult, the prevalence of these sorts of rhetorical characterizations of the "other" demonstrates the dark side of representations of disability that center its transience and the individual's triumph over it. If the ethically superior way to inhabit disability is to overcome it, it becomes all too easy to label those you deem as acting unethical as also the "wrong" sort of disabled.

Decentering Ableist Norms

In the above examples, disability exists as either an inspirational palate cleanser or as a sign of illegitimate judgment that corresponds with wrongness in body and/or mind. These representations are but explicit forms of the assumptions about whose bodies count as legitimate. Our "ableist culture's low standards for the lives of disabled people" is what makes inspiration porn inspiring, but it also

---

Lin LeMesurier, "Searching for Unseen Metic Labor in the Pussyhat Project," *Peitho* 22, no. 1 (Fall/Winter 2019); and Jennifer Lin LeMesurier, "Uptaking Race: Genre, MSG, and Chinese Dinner," *POROI* 12, no. 2 (February 2017): 1–23.

relies on defining one's place in society in terms of distance from the looming precipice of bodily failure, too often determined by a unidimensional measure of one's ability to sustain infrastructure and the economy.[40] The pandemic that began in 2020, upending people's usual ways of living, provided multiple examples of how benevolent ableism is embedded in dominant structures like health and labor. The "gratuitous violence that is ongoing and, in fact, necessary for the 'human' to continue to self-actualize without sufficient scrutiny as a category of Whiteness" that has led to the over-representation of COVID diagnoses and deaths in Black populations due to long-standing disparities in health resources and care also leads to a negation of disability as a legitimate form of inhabiting a body.[41]

Amid the pandemic, disability activists pointed out how the work-from-home situations necessitated by COVID were (a) finally providing accommodations that would support disabled workers and (b) clearly not out of the realm of possibility despite what bosses and corporations have claimed for years. These critiques focused on capitalist-focused logic that promotes maximizing work efficiency at all costs, but this logic dovetails with ableist assumptions of who counts as a legitimate worker and human. In the opinion article "Sorry but Working from Home Is Overrated" in the *New York Times*, writer Kevin Roose argues against these sorts of accommodations and warns against the dangers to workers' "creativity and innovative thinking" that working from home poses.[42] While he briefly gestures toward working from home as an option for "people with disabilities and others who aren't well served by a traditional office setup," he also argues that it is only in-person work that enables us to "express our most human qualities, like empathy and collaboration." The tacit belief that nontraditional forms of work reduce not only one's effectiveness but also one's "most human qualities" was voiced by many in arguments for a return to "normal" working conditions. Although these arguments focused on fostering empathy and creativity, rather than directly excluding disabled people, they nonetheless echo the controversial state emergency plans that explicitly categorized "persons with severe mental retardation, advanced dementia or severe traumatic brain injury" as "poor candidates for ventilator support" in the case of limited ICU supplies.[43] In these understandings of what to prioritize in the pandemic, who counts, not merely as human but as a quality human who contributes to society, is rooted in a belief that disability automatically equals less unless accompanied by overt evidence of individual compensation for this inherent lack.

The strongest portent of an ableist climate is the repeated centering of able bodies and minds as the norm and any other ways of living as not just different but worse somehow, from facile accusations of ineffectiveness to more insidious accusations of occupying a less-than-human ontology. It is this belief that disability equals less that underpins the genre of inspiration porn,[CI] hence the emphasis on overcoming and evolving past one's present disabled state. Due to both the durability of generic structures and the deeply embedded ableist understandings of living with disability as a matter of "mind over matter," it can be difficult, even for an informed viewer, to understand how these messages perpetuate harmful fetishizations of disabled actions and a generic benevolence that does not actually support needed structural and attitudinal change. Unpacking where common generic structures connect with moments of positive affect reveals knots of societal priorities that need to be unraveled in order to create lasting change in rhetorical attitudes toward disability.

The above examples demonstrate how the judgments of ability and ethics that now dog our political discourse are intertwined with judgments of capacity for rhetorical action. Accusations of mental retardation as the cause of political malfeasance are part of the same overarching orientation that poses autistic individuals as "demi-symbolic" beings, who are too immersed in their embodiment, less in discourse, and thus, in Burkean terms, as those who cannot "traverse the distance between motion and action."[44] Part of the necessary work then is developing frameworks that decenter ableist understandings of embodiment and epistemology in service of a revised definition of rhetorical agency. Nathan Stormer and Bridie McGreavy argue that we need to reconceptualize rhetoric in ways that depart from the classical emphasis on forceful persuasion, shifting

---

CI Is anyone else getting a tingle of the notorious definition of porn from the American Supreme Court Justice Potter Stewart? "I know it when I see it," he wrote, in his concurring opinion in the 1964 case, *Jacobellis v. Ohio*. You may have heard variations of the notion in the many culture wars waged since. The idea is that pornography isn't essentially definable, because it's really about crossing an invisible line, and that sort of line can be crossed in innumerable unexpected ways. All this talk of "inspiration porn" has made me wonder, what's the fundamental transgression? Jennifer's not talking about, say, "food porn," or even porn porn here, but inspiration porn seems just as challenging to the idea of norms and taken for grantedness. I think this might be the sentence where she identifies the major issue, and it's the same one we've seen in every chapter so far: the way Power uses clever strategies of naturalization to establish the other as less, while making it seem like there's no strategizing at all.

"from agency to capacity, from violence to vulnerability, and from recalcitrance to resilience."[45] Rather than assume rhetoric always emerges from an upright, fully able, ambulatory body without fatigue, fighting a climate of ableism requires looking for how "extraordinary bodies" can and do enact "inherently subversive, embodied, powerfully Other modes of persuasion."[46] Disability scholar and activist Petra Kuppers's rhizomatic framework illuminates a possible form of rhetorical agency that allows for the nuance embedded in inhabiting disability. Kuppers does not ask us to ignore moments of pride or inspiration but rather to recognize such moments as grounded in the broader epistemological diversity of different bodily experiences. Her model of disability is one

> in which the extrinsic and intrinsic mix and merge, as they do in my own physical and psychical being when I am in pain, and cannot walk up the stairs, and wish for a painkiller, and take pride in my difference (what other choice do I have?), and feel unable to speak of the nature of my discomfort, cannot find the words, but find comfort in the company of others whose pain might be different, but who somehow feel sympatico.[47]

Decentering ableism starts with recognizing how agency of *all* bodies naturally fluctuates from situation to situation in encounters with internal and external contingencies. Self-sufficiency should no longer be the primary metric for one's status as a rhetorical actor. In a heuristic that accepts disability as part of one's rhetorical capacity, there is room for the flux of embodiment that negotiates different pressures and demands through space and time.

The responsibility for rhetoricians invested in disability studies is to critically parse texts and ways of being that reproduce the existing climate of embodiment, including the definition of disability "as a solely personal phenomenon" that underpins the persuasiveness of inspiration porn.[48] By understanding how genres are capable of structuring an audience's feeling, we gain insight into how the feelings evoked by inspiration porn on an individual level are part of wide-ranging affective regimes in which ethical embodiment is defined through compensating for predetermined lacks. And these feelings are powerful; inspiration porn not only echoes other common genres like the bootstrap narrative but also promises affective fulfillment, a positive viscerality that expands outwards, making the broader political contours that demand continual self-betterment as the condition for one's existence a tad less threatening even as disability is reaffirmed as that which needs to be overcome. Parsing these dominant ways of interacting with disability is but the first step to building and supporting alternative forms of rhetorical capacity, new atmospheric understandings of the body that do not assume continuity but grapple with the multiple modes of experience that involve pain, joy, boredom and triumph, disparately.

Microclimate 3

(Jen and Bridie)

In a middle school hallway, a poster hangs on the wall. The poster shows a person in a wheelchair crossing the finish line of a race, body breaking the ribbon, hands waving in the air, broad smile on their face that also signals the pain of accomplishment. The caption asks a ubiquitous question: "What's your excuse?" Further down the hall on the door of the science classroom hangs a reproduction of *The Blue Marble* photograph of Earth, a perfectly round sphere suspended in the void of dark surrounding space. There, the caption rehearses a similar narrative: "Ours to save." Though idealized, descriptions of these types of images are not hard to imagine. These examples typify artifacts that frequently circulate within and constitute climates that continually (re)shape what counts as normal, what is seen as healthy, what is adaptive and thus desirable. We can think about such images and their modes of witnessing as "inspiration porn" and pay attention to how these images' repetitions and circulations across contexts and scales feed into and sustain climates of benevolent ableism that affect orientations toward diverse earthly bodies: human, planetary, and otherwise.

Bringing benevolent ableism and the environmental imagination together pushes us to consider how care itself, whether for Earth or vulnerable groupings on its surface, bears the potential to reinforce colonial spatiotemporalities and reinscribe harm. Habits for sensing what counts as normal and abnormal gain power within rhetorical climates, shaping an ecology of genres that render certain configurations of bodies and materials as undesirable, aberrant, and maladaptive. The belief that there are identifiable "broken" bodies that need aid from ableds mirrors the constitution of Earth as similarly fragile, broken, and in need of saving from human heroes. A climatological orientation helps show how benevolent ableism amplifies desires for able bodies and sustainable futures in related ways. For both, ways of witnessing reinforce logics of separation and objectification: the gaze is directed toward the (dis)abled bodies of humans and

Earth in ways that create a safe distance and erase the messy, heterogeneity in what a "shared future" means. These "harmful fetishizations" of bodies and futures normalize and stabilize a status quo that readily turns to technical fixes for that which is seen as aberrant, fetishizing prostheses that seek to "fix" that which isn't broken.

In climates of benevolent ableism, cognitive (dis)abilities like autism or anxiety are construed as deviant, as a barrier to *really* knowing, a condition to be overcome. To understand in this sense is a movement of an idealized body: an ability to accurately *grasp* the situation at hand as a means of control. Such ableist approaches to knowledge reinforce modes of witnessing where sustain(ability) simply becomes more of the same. The relationships among climates of ableism and environmental imagination are thus pairings that necessarily encompass sensing the intersections of personhood (as something separate from selfhood), nature, and the social. The result is an uneven balancing of what elements of life and matter we deem *significant*, and the heuristics that follow directly impact the significant's ability to exist and thrive.

What would differential witnessing of Earth and bodies entail? Such inhabitation of alternative alignments enables a fuller recognition of how "orientations, which come out of obligations, mean that you are facing a particular direction with a specific horizon of possible action before you. Orientations are the condition of possibility for some futures and not others."[1] While climates might evoke beliefs/actions in individual rhetors in real time, they also maintain a sort of threshold level of said beliefs and appropriate actions as always available for rhetorical use. To think with climates, as well as with affects and ethics, is to think with different possible futures, shifting who or what is significant. Taking such an approach to rhetoric destabilizes the demand for timely, linear responses from an upright position. Instead, the ability to "digest truth through body, to soil" is centered as a way through to new insight.[2] But "use" can be a form of violence inimical to listening—a posture of orientation already appropriated to serve a predetermined end. Floating, on tides, rejiggers what urgency feels like, what embodiments are not only possible but also held as epistemologically valid or legible.

4 Disrupting Environmental Imagination, toward a Tidal Ethics

(Bridie)

Imagining the earth from space is an ancient practice, one that the science of space travel radically expands. Of the suite of images of earth from space produced through time, none have gathered more rhetorical force than *Earthrise* and *The Blue Marble* (figure 1). As Sheila Jasanoff argues, images "of the earth suspended in a *void*, captured by cameras of the U.S. space program beginning with the Lunar Orbiter of 1966 and culminating with the Apollo 17" constituted "a global environmental consciousness" promising "an imagined community as encompassing as the earth itself."[1] Of this promise, Jasanoff also asks if "those without the power to patrol the heavens, to map and perhaps to devastate the earth, can ever meaningfully participate?"[2] Though she poses this important question, her analysis does not engage it. Instead, she focuses on the limitations of the argument that images of earth from space alone contributed to radical shifts in environmental consciousness to show how "images become persuasive only when ways of looking at them have been carefully prepared in advance."[3] This chapter picks up where her analysis leaves off—and also builds out from the chapters by Chris, Nate, and Jennifer—to more fully consider the exclusions that are reproduced in the consistently affirmative assertions that images of earth from space inspired the environmental movement and, thus, greater care for our home planet. In contrast, this analysis turns attention to how images of earth from space rather feed into an environmental imagination shaped by logics of temporality, categorization, and territorialization—constituting both "the environment" and what it means to be human in racialized and colonial ways. In other words, in the context of a global commitment to the science

FIGURE 1. *Earthrise* and *The Blue Marble.*

of space travel, images of earth from space serve as powerful objects through which racialized and colonial orders are coproduced.

Though focusing on images of earth from space offers a way to illustrate logics that guide environmental imagination, the questions that motivate this work come from a different and distinctly earth-bound space, namely community-based collaborations focused on restoring intertidal mudflats. My experiences with intertidal restoration are shaped by environmental imagination in ways that produce continuous negotiations, contradictions, and complicities, as the following brief example intends to illustrate. In the place I live, the homelands of the Penobscot and Wabanaki Nations in a region also called Maine, intertidal mudflats are governed through a municipal shellfish comanagement system. The legal standing for this arrangement was established in 1641 through the Massachusetts Bay Colony Ordinance in a document known as the Body of Liberties. This colonial document grants rights to settlers to access intertidal mudflats for fishing, fowling, and navigation and enables coastal municipalities to make decisions about how to manage intertidal resources like clams, mussels, and oysters.[4] This local governance context enables communities to negotiate and challenge state power. However, the law also creates a system through which intertidal spaces are treated as municipal property, a process Édouard Glissant describes as territorialization,[5] which serves as a formative logic that then shapes what "the environment" becomes with consequences for, in this case, coastal inhabitants. This territorialization is a structuring process where those who do not belong to a municipality are excluded from accessing this space, including Wabanaki peoples who have long relied on and stewarded intertidal shellfish. "Environmental imagination" names a way of making sense of logics, like territorialization, that guide racialized and colonial social orders, like how shellfish comanagement contributes to land dispossession. Feeling the constraints of this way of relating to intertidal mudflats, as well as sensing how the tides themselves push on and resist stable definitions of land and territory, I increasingly felt the need for alternatives to disrupt this formation and turned to theories shaped by tidal forces.

This chapter suggests that a tidally influenced approach to knowledge can destabilize dominant, racialized logics and help, as Tiffany Lethabo King says, "sculpt new epistemologies and sensibilities that shape the contours of humanness in more expansive ways."[6] Reconsidering images of earth from space serves as an initial step in developing a visual rhetorical methodology to enable critical, affective, and embodied ways of tracing images. The second step involves bringing this methodology to an analysis of how images of earth from space articulate and reinforce logics of time, categorization, and territory—often in service of a particular but exclusionary environmental imagination. Sketching some of the features of environmental imagination helps reveal, by contrast, how the earth's tidal forces produce exigencies for critical praxis that can challenge the totalizing logics that constitute the earth as unified, singular, and fixed. Moving toward tidal ethics aligns with one of the core tenets of this book, and of our reading group's playful thinking-feeling together over the last few years: namely, the sense that relational, listening-based approaches to knowledge coproduction[CR] offer generative ways to "better understand our action in the world."[7]

CR I really love this. I wonder here what kinds of receptors and capacities you've had to build to engage in work oriented toward tidal ethics. I seem to constantly have my mind wrapped up in questions of pedagogy these days, maybe, but I wonder: What conditioning to the climate and to those within it are required to build capacities for relational listening, thinking-feeling, playfulness, and the coproduction of knowledge? What does it take to build these capacities or make them more likely to appear in the ecology, both within individuals and also within communities, institutions, and climates?

Contextualizing Images of Earth from Space

In his book, *Earthrise: How Man First Saw Earth*, Robert Poole describes the cultural impacts of *Earthrise*, captured during the 1968 Apollo 8 mission, and *The Blue Marble*, taken during the 1972 Apollo 17 mission. The following description exemplifies dominant characterizations of the relationship between earth images from space and the emergence of environmentalism in popular texts and news media:

> The space programme changed thinking about the Earth, but not in the way that either its supporters or its critics expected. The Apollo years of 1968–1972 coincided almost exactly with the take-off of the environmental movement. *Earthrise* was followed by Earth Day. As men journeyed from the Earth to the Moon, the human race made the philosophical leap from Spaceship Earth to Mother Earth.[8]

This quote connects with a persistent pattern in interpretations of these moments of witnessing during the Apollo space program, which emphasize the positive social and cultural influences associated with visualizing earth from space. The dominant interpretation of the impact of seeing images of the whole earth from space contends that these images allowed humans, for the first time, to imagine themselves on a shared, fragile, and singular planet.[CI] This form of witnessing is akin to Jennifer's notion of "inspiration porn," explored in the chapter before this one, where the earth itself is apprehended as an idealized

---

CI Of course, neither *Earthrise* nor *The Blue Marble* show us the whole earth, only one vantage on it, leaving the dark side to the imagination. But these photographs do show the planet as an imagined whole, that is, from sufficient distance as to identify a separation between planet and not-planet. Aerial photography of earth has been around since the advent of photography itself; it wasn't until advanced space travel that any cameras were distant enough from earth to portray that separation. In this sense, the salience of these images isn't just to visualize Earth, but to visualize the planet within a context of what exceeds the planet. The lesson being: Earth is not the end; how small we are. Or, as Marshall McLuhan put it, "'Ecological' thinking became inevitable as soon as the planet moved up into the status of a work of art." See Marshall McLuhan, "At the Moment of Sputnik the Planet Became a Global Theater in Which There Are No Spectators but Only Actors," *Journal of Communication* (Winter 1974): 49.

body. Further, these reflections often comment on how a mission to better understand our place in the cosmos turned into a journey to rediscover the earth and our stewardship responsibilities. In celebrating the fiftieth anniversary of *Earthrise* in 2018, a *Guardian* article rehearses this narrative:

> Before that moment 50 years ago, no one had seen an earthrise. . . . What [Bill Anders] captured became one of the most influential images in history. A driving force of the environmental movement, the picture, which became known as *Earthrise*, showed the world as a singular, fragile, oasis.[9]

Narratives frequently emphasize how earth images coincided, and in some cases, initiated the environmental movement, as if the images themselves were responsible for the emergence of a distinctly environmental consciousness. What these interpretations generally fail to convey is how *Earthrise*, *The Blue Marble*, and the suite of images produced during these early days of space travel materialized within a context that was already ripe for the kinds of articulations that these images have since reproduced.[10]

Though these images certainly coincided with the environmental movement, there was not a singular driving force in creating environmentalism. Instead, these images emerged within a context in which scientific endeavors and cultural values that shaped relationships with the environment were already deeply enmeshed, creating the climate in which these images came to circulate. It was the very relationships between science and culture that helped to create the conditions of possibility for *Earthrise* to gather persuasive force.[11] These relationships go well beyond the material conditions of astronauts in a spacecraft circling the moon to include a series of institutionalized practices that put a camera in astronauts' hands and, more broadly, aimed to put astronauts in space to satisfy nationalist desires.[12] For example, Poole provides in-depth detail about the behind-the-scenes negotiations about when, how, and by whom photos of earth should be taken. All of these negotiations were connected with internal NASA policies and publicity plans with the objective of producing images that would help persuade the American public about the value of space exploration. Beyond NASA as an institutional context, these images took on persuasive force through international relations that comprised Cold War politics and global competition to be "pioneers" of space.

Being the first in space was a matter of national pride, effectively connecting images of the earth with nationalistic identity and complicating, if not contradicting, the argument that "the mightiest shot [of *Earthrise*] in the Cold War turned into the twentieth century's ultimate utopian moment."[13] In contrast, as Jasanoff argues, "While astronauts, astronomers, and international experts identified the earth image with coexistence and political interdependence, the use and enjoyment of environmental resources remained for many other actors tightly bound to national interests."[14] More complete histories of earth images have been reviewed elsewhere, but this brief discussion begins to show how images do not simply appear out of a "void" where "darkness was upon the face of the deep" and how images and imagination are mutually constituted through rhetoric.[15]

A Visual Rhetorical Methodology

Rhetoricians have long been interested in how images circulate to shape social orders and constructions of the environment. This scholarship has contributed important insights about the relationship between images, materiality, affect, articulation, and arrangement/order. A rhetorical climatological orientation builds from these insights to amplify attention to how these types of influences relate to formations of power and oppression. For example, Kevin Deluca and Anne Demo examine how Carleton Watkins's landscape photography of Yosemite in 1861 serves as "founding texts in the construction of a wilderness vision that has shaped the contours and trajectory of environmental politics."[16] As with images of earth from space, Watkins's representations emerged within cultural contexts that preceded them, such that these "images not only resonate with, but also comment on, larger cultural narratives regarding national identity, scientific and industrial progress, and even race and class privilege."[17] These images also intensified ongoing forces of physical and cultural genocide:

> Just ten years prior to Watkins' discovery of Yosemite, the Mariposa Battalion had entered the valley with the intent of relocating or exterminating the Ahwahneechee. The ability of whites to rhapsodize about Yosemite as paradise, the original Garden of Eden, depended on the forced removal and forgetting of the indigenous inhabitants of the area.[18]

Images articulate wilderness with environmentalism in ways that reinforce socially constructed binaries between nature and culture and violent erasures of Indigenous inhabitation and land relations. As Max Liboiron puts it, "Environmentalism does not usually address and often reproduces colonialism."[19] Thus images constitute environmentalism in ways that are entangled with ongoing forces of oppression, including racism, colonialism, and genocide.

How, then, does a rhetorical climatological approach to images shape ways of tracing relationships between images, environmental imagination, and forces of oppression? Tracing visual rhetorics means attending to the ways things get linked together, or the fine-grained practices of articulation that constitute orders of matter, meaning, and power. Nathan Stormer draws from the rhetorical concept of *taxis* where articulation is "to do order and be done by order."[20] As applied to images, articulation means following modes of arrangement and how images become linked with texts, networks, and within rhetorical climates of meaning and feeling, where the suasiveness of an image is shaped by "the interrelationships articulated by practices."[21] In a similar vein, Laurie Gries focuses on articulation and circulation of images within networks as she traces myriad manifestations of the Obama *Hope* poster. Her visual rhetorical methodology enacts "a spatiotemporal, distributed process that intensifies with each new actualization and with each new encounter."[22] In terms of environmentalism, Kevin Deluca also draws from articulation theory to describe environmental activists' practices to create image events, affective encounters that circulate through mediated systems of representation and shape environmental activism.[23] A rhetorical climatological orientation to images thus takes a nonlinear, circulatory approach to attend to diverse practices that link things together, such as connecting earth images from space with a sense of planetary unity with a shared experience of humanity with environmentalism. The method also attends to linkages that amplify affects, or how images intensify emotions like awe, wonder, and horror.

Further, turning to race, gender, embodiment, and ableism as analytics and not as subjectivities or identities draws attention to intersectional structuring forces of oppression. A focus on articulation allows tracing how categorizations become ordered as hierarchies through processes of normalization where some bodies and languages are privileged "while others are not. A central part of that normalization of order is to establish the parameters of possible interaction among the various elements being normalized."[24] For example, Kim TallBear's analysis of genomic science, Native American race, and tribal enrollment processes as racialized governance regimes identifies how practices in science rely on genetic or species-focused categorizations to coproduce racialized social orders.[25] Histories that articulate gene as DNA create the parameter conditions through which scientific processes of categorizing species based on genetic difference fold into categorizations of and hierarchies among human races, and images figure centrally in this process. As Zakiyyah Iman Jackson argues, the production of species as a scientific object was, and continues to be, apprehended as a matrix in which "racial slavery, conquest, and colonial ideas about gender, sexuality, and 'nature,' more generally, have informed *evolutionary discourses on the origin of life itself*."[26] Jackson names this matrix "(anti)blackness" which constitutes:

> a mold, a womb, a binding substance, a network of intersections, functioning as an encoder and decoder. It is an essential enabling condition for something of, but distinguishable from, its source—and therefore, it performs a kind of natality, performing a generative function rather than serving as an identity.[27]

When (anti-)Blackness is situated as matrix or rhetorical climate, visual articulations of genes and, by extension, earth images from space take on distinctly different significance, especially in contexts where these images intersect with the globalizing force of science. A climatological orientation to race, and especially the constitutive relationship between Whiteness and (anti-) Blackness, requires returning to commonplaces, like the environment, human, climate change, Anthropocene, and so forth, to consider the racialized histories that shape the social orders that these concepts articulate and coproduce.[28]

Such analyses also begin to point to how science *should* be conducted to challenge racialized and related social hierarchies. Knowledge coproduction offers a complementary theoretical orientation to articulation theory in its attention to the mutual constitutive effects of science and social orders. In the context of collaborative approaches to science, an orientation to knowledge coproduction can produce a set of research commitments that attend to the mutual constitution of science and society to alter patterns of dominance in science that reproduce societal inequities.[29] Approached in this way, knowledge coproduction can become a process of attempting to enact a grounded, contingent, and emergent ethicality in and through science, a commitment to which I return in the concluding section.[JL]

In sum, images are rhetorically powerful for how they articulate matter and systems of meaning and power, and for how they reproduce racialized, colonial, and ableist relationships with the earth and one another. Imagination and knowledge coproduction offer conceptual and practical apparatuses for attending to these articulations, as these concepts hone attention to the mutual constitution of science and social orders such as state power, colonialism, race, gender, and (dis)ability. Thus, the analysis that follows reconsiders a host of articulations through which earth images constitute environmental imagination as climates of intersecting forces of oppression.

Articulating Environmental Imagination and Crisis

There are a number of influential media texts that show how images of earth from space contribute to a distinct sense of environmentalism. One of the most common representations, iterated in various ways, depicts an *Earthrise*-esque image of the whole earth held by one or two, usually White, hands. The representative example provided in figure 2 depicts the pattern where earth images are articulated with negative human impacts on the environment that have created an urgent and unprecedented crisis, connections that draw from and simultaneously reconstruct a "crisis epistemology."[30] This rhetoric of crisis is a persistent pattern in the constellation of meanings that circulate with earth images and can be traced in arguments associated with the emergence of the environmental movement in the 1970s, where crises of environmental contamination, species extinction, and natural resource shortages were central concerns and where climate change as crisis was beginning to take hold.[31] Framing climate change as an urgent and unprecedented crisis serves as a central and recurrent theme in climate change discourse where visual images, like the iconic hockey stick graph, serve as visual rhetorical strategies through which "science places

JL It strikes me that such an approach is important in how it foregrounds the rhythm of care that is necessary to foster an ongoing, "emergent ethicality" that is reciprocal rather than reactionary.

vision and sight as the primary sense for achieving objectivity," thereby feeding into constructions of "*unprecedentedness* and *urgency*" as central features of a crisis epistemology.[32] In arguing for a shift from a crisis epistemology to one of coordination, the latter of which would "organize knowledge through the vector of kinship relationships," Kyle Whyte sketches the contours of a crisis epistemology:

> There is a complexity or originality to the imminent events that suggests the need to be immediately solutions-oriented in a way believed to differ from how solutions were designed and enacted previously . . . [whereby] time is put together (arranged) to favor a certain conception of the present as a means of achieving power or protecting privilege.[33]

Whyte's characterization of the arrangement of time points to the role of memory in articulating a present in ways that reinforce strategic forgetting of both histories of oppression and the existence of multiple temporalities. The emphasis on the unprecedented event of climate change remembers the present in ways that require forgetting how large-scale and socially devastating global changes have occurred and continue to persist.[CR]

Further, urgency remembers temporality in discernible ways,[JL] as urgency connects with and reinforces linear, neoliberal, and colonial temporalities.[34] These temporalities are shaped by progress narratives, logics of hierarchy and control, and related paradigms of pioneerism and frontierism, all of which are central to colonial formations of power.[35] A clear form of evidence for this pattern in space travel is the frequency with which these missions are described as "pioneering" and aiming to cross the final frontier. For example, commentators frequently

---

CR I am thinking here (and with your conversation immediately below on how coloniality presents time as linear and within progress narratives) about what it might mean to imagine a critical praxis that centers time. For example, what would it mean to labor to keep alternative conceptions of time present and alive? How might we carve space for such times? Or pursue such time in our spaces? Would this be a labor of critical time management, managing of time? How is the practice of disrupting the smooth time of coloniality tied to tidal ethics?

JL Your phrasing here gets at the tension between "discernible" and "remembering," underpinned by the verb form of "discerning." These rememberings and how they make time legible are also forgettings of other ways to slice up time, but those ways of forgetting then become elevated as purer ways of being in the world.

FIGURE 2. Screenshot from a PBS LearningMedia website that organizes education materials, showing a stylized yet iconic rendering of an earth image from space. PBS LearningMedia, "Human Impact on the Environment."

celebrate and articulate the Apollo missions with Christopher Columbus's genocidal mission to what is now known as the North American continent.[36] In a press conference shortly after Apollo 8 landed, Thomas O. Paine, NASA's then chief, compared the success of the Apollo 8 mission to Columbus's voyage,[CI] and this comparison became a repeated refrain in how the mission was remembered across texts.[37] Further, when Frank Borman visited Europe to meet with celebrated leaders in the months following the Apollo 8 mission, he made a special stop in Madrid to lay a wreath on the statue of Columbus. Poole points to this performance, the wreath laying as well as the broader tour, as an example of how "everywhere [Borman] took the same 'one world' message about the need for peaceful coexistence and international brotherhood," a message distinctly at odds with his memorialization of Columbus's achievements.[38]

Thus, the kairos of urgency promotes a forgetting of multiple temporalities

CI Oh goodness, yes. The repetitions of history that don't acknowledge (and worse: glorify) atrocities of the past are so horrible. A relative of ours who works in the aerospace industry once gifted our son a T-shirt that says "Colonize Mars" in big bold letters across the chest. I've refused to let him wear it. But it's also inspired positive conversations, albeit some I hadn't imagined having with a four-year-old. The imperilment of our planet, though, does at least raise some questions about whether colonization requires people to be colonized.

as well as a forgetting of the multiple logics that trace complex problems like "climate change." For example, climate change can be remembered not as an outcome of fossil fuel emissions but instead an outcome of the logics of exploitation, founded in the practices of slavery and colonialism, that constitute dominant relations between humans and with the earth.[39] Forgetting these more complex patterns is requisite for the articulation of technical solutions and for working with "stakeholders" and "decision makers" for whom such solutions would be applicable and desirable. In related work, we attend to the relationships between constructions of time and "sustainability solutions" in collaborative, knowledge coproduction research that addresses climate change and environmental issues. In our experience, rhetorics of urgency associated with grant-funded research reinforce neoliberal logics and colonial research practices as well as existing power disparities. In the context of climate-focused and environmental science-based collaborations with Wabanaki peoples, attending to multiple and Indigenous approaches to time, such as tracing time with the rhythms of the Penobscot River and through fish migration, helps challenge these logics and change patterns through which science and social orders are coproduced.[40]

Memory and Becoming Human

Linear, clock-based formations of time that have come to govern life on earth serve as an ontological marching drum of environmental imagination, with its neoliberal rhythms of progress, universalism, and control.[41] These rhythms are sustained by material practices that (re)construct what it means to be "human" in at least two ways. First, remembering images of earth from space as representations of a fundamental connectedness to the earth enacts a forgetting of the ecological conditions of their very production.[JA] Said another way, physically witnessing an image of earth from space relies on the most profound material disconnection from earth that humans have ever experienced. Without their own "spaceship earth," a wholly manufactured

JA Waiting for the mud in tidal pools, the "rhythms of the Penobscot River," for there too lies the ossuary. Mud to the moon and back again. I so hope the NASA (and NOAA) lenses continue to be refined, to show from afar the drought and poverty, the racially and ethnically divided, the impoverished territories and such "shoals" from T. L. King.

environment that partially replicates earth's life support systems, direct witnessing of earth from space as astronauts is materially impossible. The emphasis on human witnessing as an embodied performance matters too, as the ways in which Anders, Borman, and Lovell experienced *Earthrise* themselves, reading from the book of Genesis in what was at the time the most widely watched television broadcast ever produced, and then touring the world talking about this moment for years afterwards intensified *Earthrise*'s persuasive force. It matters that the astronauts were on the spacecraft, lost radio contact, and then captured the image just as they regained contact. It matters too that in one of the more iconic renderings of this moment in modern environmentalism, namely from the opening scenes from *An Inconvenient Truth*, Al Gore misremembers this moment. The slippage and inaccuracy in Gore's account helps illustrate the constitutive effects of this image. In contrast to his narration in the opening scenes of this film, the astronauts did not have radio contact when they captured *Earthrise*. It was only from the recording on the craft itself that people have been able to listen to that embodied moment, a scene that is featured as the opening of Poole's book and described in detail in countless news articles.[42]

What is this slippage about? This slippage points to how embodiment and imagined subjectivity come to matter in how these images shape meanings of "the human." As Poole notes, "The first photograph of the whole Earth was taken not by astronauts but by an ingenious orbiting photographic laboratory called Lunar Orbiter."[43] In a more contemporary example that has also been largely ignored in news media, NASA recently reproduced an *Earthrise*-like image using a robot satellite from the surface of the moon. The image is an eerie composite of *Earthrise* and *The Blue Marble*, made even stranger by how it puts humans in the position of witnessing earth from space from the standpoint of a robot on the moon.[44] The earliest and subsequent suite of images of earth from space produced by robots have not circulated with the same kind of rhetorical force as images witnessed and captured by humans. These material conditions and embodied/technological performances are not ancillary to the affective, rhetorical force of these images. Instead, by paying attention to the affective and ambient, yet always materially embodied nature of rhetoric, a climatological orientation contributes ways of tracing how complex ecologies, in this case ecologies associated with life on a spaceship, contribute to specific articulations of images of earth from space that shape meanings of both "the environment" and "the human." In this case, embodied witnessing creates a tension between the profound disconnection from earth's life support systems and shared identifications with this view of the earth. These tensions are intensified within a scientific context that invites us to imagine that if humans can do this (i.e., survive without the earth), "we" can, in fact, do anything.

In the case of *Earthrise*, remembering a profound moment of connection to the earth requires an equally profound forgetting of our reliance on earth's life support systems. Extending this point further, every time *Earthrise* is remembered as constituting humanity, that sense of humanity is only made possible by being materially disconnected from earth. Though there are many practices that articulate this particular tension, the physical act of raising a camera to *capture* this image transforms a moment of radical, sensuous, and multidimensional experience even in the context of a spaceship "toward two-dimensional visuality" and further shows how "Western vision sensory dominance relegates aspects of nature to a safe distance that renders them passive and contracts embodied experience."[45] Here Tema Milstein points to the constellation of visual practices that contributed to how *Earthrise* and *The Blue Marble* simultaneously rely on and intensify logics of separation, hierarchy, and control.

The transformation in meanings associated with naming these images also signal the second way in which these images constitute "the human." It is significant that *Earthrise* first names an earth that can rise on its own, and then, in contrast, *The Blue Marble* names a relatively inert object that fits in a human palm and that requires human ingenuity to propel it. The concept of a marble corresponds with a frequent visual trope associated with "the environment," visualized as (White) human hands holding a small, round earth, depicted in the PBS image above and in many related images that are frequently used in representations of environmentalism. Rhetorical analyses of environmental images like these tend to focus on how naming practices feed into or challenge dominant power, such as in anthropocentric ideologies, and (re)constitute binaries such as human–animal and nature–culture. Yet, such analyses accept categorizations, including the *anthropos* in anthropocentrism and the human and animal that stand on either side of the binary, as given. Focusing on the dialectics, but taking for granted categories such as "the human," contributes to epistemic erasures of how race as a process of categorization "provide[s] a vehicle for reinforcing a striated conception of human species,"[46] naming certain humans with specific abilities to hold, grasp, and control. Jackson calls attention to how racism haunts concepts like "human" and "animal" and argues that these categories come to make sense within a matrix of (anti-)Blackness. Awareness of how racism articulates with concepts like species, humans, animals, nature, culture, gene, plastics, environment, and so forth "raises the stakes" of critiques of dialectics because these "studied critiques of nature–culture oppositions and the phenomenon itself are inside of the economies of affect and desire generated by raciality,"[47] where the racialized context of scientific discourse functions as "an essential means of arranging human–animal and human–nonhuman distinctions."[48]

How does a racial analytic unsettle the categorizations that have shaped much of the scholarship on anthropocentrism and dialectical modes of relationality with the earth? To begin to address this question, we can return to the example of the (White) hands *grasping* the earth above (figure 2). In contrast to approaching race as identity, which would read Whiteness visually encoded in these hands, the action of grasping is onto-epistemologically significant for what this image comes to mean, for the subsequent orders that this image performs, for the climates that are (re)produced. Jackson calls attention to this when she draws from Glissant who "contends that when considering the development of Western imperialism—from 'discovery' and territorial expansion to anthropological ethnography—the verb *to understand* in the sense of '"to grasp" [*comprendre*] has a fearsome repressive meaning.'"[49] The visual order and performance of the grasping hands, one that is evident not only in the example in figure 2 but also in the nearly ubiquitous way in which earth images from space have come to be articulated with environmentalism, reify a relationality of *knowability*: the earth exists to be described, measured, understood, and held in full view by humans and where (anti-)Blackness orders the hierarchies of *which* humans and *which* forms of knowledge are assumed in these relations.

Calling attention to categorization as a mode of racialization that orders binaries, such as human–animal, nature–culture, or many others, challenges new materialist and ecological rhetorical approaches that seek to transcend or blur the boundaries between human and nonhuman entities or that argue for rhetorical strategies like granting personhood for nature. Instead, a racial analytic traces how "logics of race are determinate of logics of species, and ecologies of antiblackness shape epistemologies of scientific thought and their taxonomies that purport to divide the human from animal."[50] Further, though the analysis in this section emphasizes racialization and climates of anti-Blackness, one can also trace how ableist climates seep into this formation, where images of earth from space construct knowledge through idealized embodied activities of grasping, viewing, and rationalistic description. This analytic thus interweaves attention to (anti-)Blackness with epistemologies of time, science, and visual practices to begin to account for how the environment and its attendant terms, like the human, are shaped within intersecting oppressive climates. This orientation thus shows how *Earthrise* and *The Blue Marble* intensify racialized patterns of dominance and hierarchy in and through the science of space travel as a "global" knowledge project.

From Void to Territory, (Anti-)Blackness to Environmentalism

The preceding sections established how the science of space travel and images of earth from space coproduce linear and colonial formations of time, and especially epistemologies of crisis.[NS] These temporalities enact a forgetting of the multiplicity of time and the violence of racialized oppression. The argument that changes wrought by climate change are unprecedented forgets the devastation of slavery and colonization and how climate change is an ongoing manifestation of rhetorical climates that abide logics of territory and exploitation and not, more simply, a result of burning fossil fuels.[51] Further, visual witnessing of earth from space reproduces affective tensions that intensify these modes of forgetting and, in particular, forgetting intimate material connections with the earth as the basis of existence. Turning to Jackson's argument about (anti-)Blackness helps push on scholarship that would interpret these tensions and modes of forgetting in terms of human–nature dialectics and the formation of anthropocentrism to instead remember how categorizing "the human" occurs through a climate of discursive practices of racialization with partial origins in the development of the concepts of "species" and "gene" in evolutionary science. In this section, I continue the analysis of how (anti-)Blackness serves as matrix, the defining parameter through which subsequent social orders emerge, to reconsider the epideictic performances

---

NS I so appreciate your analysis of these images because it helps one understand that there is no innocent, uninflected visualization of the whole of "us." The technical prowess and the beauty of the earth makes that hard to appreciate, providing alibis, I think, for the temporal and colonial visions manifested. Notions of "world" always form a place in relation to totality, as Glissant says. I think you really make that understandable here.

of scripture and poetry that circulate with *Earthrise*. This approach intends to show how (anti-)Blackness is also gendered and how, in the context of images of earth from space, this becomes evident in the frequent articulations with "the void." This gendering is also sustained through constructions of linear time and territory, and here Glissant's work gives language for how these constructions feed into the formation of environmental imagination.

When Apollo 8 astronauts reestablished radio contact on Christmas Eve 1968, audiences watched a grainy image of the lunar sunrise and listened to a short reading from the Book of Genesis, as Anders, Lovell, and Borman took turns reciting lines like "In the beginning, God created the heaven and the earth. And the earth was without form, and void; and darkness was on the face of the deep" and "And God divided the light from the darkness," and then concluding by wishing all "good night, good luck, a Merry Christmas—and God bless all of you, all of you on the good Earth."[52] While the spacecraft was making its way back, and before audiences on earth had seen the photograph of *Earthrise*, Archibald MacLeish published a poem that was then frequently reproduced with the image. Echoing the meanings conveyed in the Book of Genesis reading, in particular a distinct sense of brotherhood, fellowship, and unity, MacLeish concludes his poem:

> To see the earth as it truly is, small and blue and beautiful in that eternal silence where it floats, is to see ourselves as riders on the earth together, brothers on that bright loveliness in the eternal cold—brothers who know now they are truly brothers.[53]

In contrast to the homogenizing interpretation that images of earth from space construct a singular and shared sense of humanity, these passages point to the broader rhetorical climate in which these images reinforce regimes of racialization that depend on (anti-)Blackness to order social hierarchies that are simultaneously gendered. On one level, the metaphors of (anti-)Blackness, and especially the recurrent and racialized tropes of light and dark, are immediately apparent. Yet, similar to the limitations of a social constructivist or representationalist orientation noted in the above analysis of human–nature dialectics and anthropocentrism, emphasizing the constitutive effects of these representations runs the risk of obscuring the ontologizing function of anti-Black *femaleness*, especially the implications of frequent articulations with the "void."[CI]

As Jackson describes, comparisons "of blackness's resemblance to nothingness reveals an anxiety about a declension into a void. Attempts to nullify blackness has a sexuating logic . . . one that figures black(ened) femaleness and/or femininity as baleful, phobogenic fleshy metaphors of the void."[54] Examples and quotes used throughout this essay, including in the opening paragraph, show how images of earth from space are frequently described as appearing out of or suspended within a void. In this sense, void exceeds the symbolic inscription or the figure "void" to also include a series of "fleshy metaphors" that amplify the sense of anxiety that Jackson describes, such as how this apprehension of earth and our shared humanity was birthed from an "eternal silence" and "eternal cold" in MacLeish's verse.

This anxiety becomes a motive force for the interlinked impulses of science, territorialization, and the myth of brotherhood (i.e., filiation) to constitute humanity as Whiteness, which relies on (anti-)Blackness for its formation—as Nate shows in chapter 2. In terms of this constellation of forces, Glissant argues that dominant science and territorial conquest are interchangeable commitments, both profoundly shaped by how Christian narratives of brotherhood construct a sense of relatedness (filiation as knowing our past) and, simultaneously, articulate a linear, descendent-based temporality with knowledge. Comparing science and filiation, Glissant says, "Both are linked to an identical spirit of universality—in opposition to community's exclusivity or to Nature's heterogeneity; both reach fulfillment at the end of a line; both already are and further become the propagation of a Knowledge."[55] Approached in this way, science and filiation are interlinked forces that spur urgent conquests of territory and for knowledge. For example, when Apollo 8 astronauts read from Christian scripture to invoke brotherhood as an attempt to communicate care for the future of this shared planet, they draw from the past to do so and construct time as a linear sequence. Further, the myriad articulations of *Earthrise* as image, the Book of Genesis, and MacLeish's poem serve as further evidence

CI Bridie has such a keen ear for the ways certain words carry insidious connotations. But what's so amazing is that she pulls on those words until she draws innumerable related oppressions out of their hat. Maybe without even quite trying to, this chapter is teaching me that intersectionality is not just about identity. Oppression and exclusion are themselves intersectional magnets: they draw everything subordinate into their own voiding "void."

for the complex relationships between Christian narratives and science. In contrast to what brotherhood represents when performed from the subject position of White male astronauts reading Christian scripture, attending to the ontological dimensions of brotherhood can help orient toward the logics through which filiation articulates humanity as Whiteness, the racialized and gendered power assumed in the choice to become "riders on the earth together." Just as imagining the earth from space occurred within scientific and cultural climates that prepared audiences for this visual practice, Western mythologies, like the story of Genesis, "[universalize] linear time—before and after Christ" and bring "a chronology of the human race into general use," including in science, in ways that temporalize, categorize, and racialize "the human."[56]

Toward a Tidal Ethics

Focusing on how images coproduced within the science of space travel constitute environmental imagination as a rhetorical climate helps illustrate what Jackson is getting at when she describes, how "a transculturally verifiable image of the earth, or positivist knowledge as aspirational horizon, has been pursued via a combination of material-discursive force and a coercive (dis) possession of processes of sense perception and cognition on a global scale."[57] Apprehending these images through a racial analytic like Jackson's, (anti-) Blackness as a dominant rhetorical climate underscores how articulating these images with environmental consciousness merits critical attention. But that attention must also consider what it means to be human, especially because epideictic rhetorics that celebrate the positive value of these and other images tend to circulate with such intense frequency and durability. In contrast to dominant narratives that articulate these images with a positive vision of environmentalism based on a global unity, shared sense of humanity, and care for the earth, these images instead intersect with the globalized science of space travel in ways that adhere to and reinforce logics of time, categorization, and territorialization to constitute environmental imagination as a racialized and colonial climate that, unfortunately, still endures. Its endurance raises an important question: what are possible pathways to an alternative climate, an alternative imagination that does not rely on dominant science or sense dispossession as its parameters of possibility?[NS]

Glissant's suggestion that "we must also involve this imaginary in the place we live" offers one potential response.[58] And it might draw us toward the ocean tides.[59] Much of "my work," like this book itself, is always work-with-others insofar as it treats knowledge as relational, community-based, and coproduced. Doing so, for instance, in the context of intertidal mudflats, necessarily attends to how rhetoric shapes initiatives to restore and sustain these places with those communities impacted by their harm. These collaborations have identified commitments that allow us to begin to address structural inequities that were created through colonial and neoliberal forces that sustain and intensify Indigenous land dispossession and rural poverty.[60] The way in which we shape our research intends to find ways to unclench the grasp of dominant logics of science to instead situate this work in/as relation. As a theoretical-ethicomethodological commitment, knowledge coproduction organizes research praxis in ways that acknowledge that how we produce knowledge constitutes worlds, or articulates orders of matter and meaning. Further, coproduction recognizes that while some emerging orders can be followed and traced, the world and its myriad elements are not ultimately knowable in a final sense. Instead, a relational approach to knowledge attunes to the poetics of worldmaking, relying on affective sensing, movement, and description as the basis for making ethical choices. Instead of trying to define, name, and generalize a singular tidal imaginary, this orientation tries to work with incommensurabilities, recognizing that "opacities can coexist and converge, weaving fabrics. To understand these truly one must focus on the texture of the weave and not on the nature of its components."[61] As chapters in this book help show, an orientation to rhetorical climatology is about feeling and following the threads and contours of this weave.

NS One of the questions that your work always prompts in me is how to address the imaginary in a place-based way such that tidal ethics does not turn using tides and shorelines to think with into a simple metaphoric exercise. What about the dynamism of tides, as ways of inhabiting permanence and change, can change our orientation? E. Cram talks about queer orientation and uses orientation as a way to address ecological inhabitation. For me, I think the way that urban living so powerfully orients rhetoric can be a useful contrast for thinking tidally in ways that are more than a metaphor as people often like to think of metaphors. The blue marble is seen from a largely urban orientation, and tides are part of the picture you see from above. If tides and tidal places provided an orientation, which would itself queer the urban, that could be productive.

FIGURE 3. This is a photo from a collaborative effort to restore an intertidal mudflat using "brushing." Brushing involves planting twigs and branches in the mud, both to help catch clam seed as it washes in on the tide and to help the seed settle into the substrate. Photo by Bridie McGreavy.

This weave can be visualized in a photograph (figure 3) of an intertidal conservation event in which clammers and volunteers planted brush in the mud as a means to draw clam seed down into the mud and grow clams.[JA] As apprehended within environmental imagination, brushing is futile, "scientifically demonstrated" to have no effect on growing clams. As constituted through listening and tidal ethics, brushing instead becomes a means of standing with each other within mudflats and making poetic choices amid limitless uncertainty and difference.[62]

In our community-based work, naming environmental imagination has emerged as a way of attending to the texture of the weave—matrix, climate,

JA This image evoked in me a former and familiar notion in conversation, Bridie. No tidal pools in the Rockies, but one can stand in the mud of the Platte River after the flood, or stand in darkened snow from the soot of fires. Tidal ethics above five thousand feet.

imagination—for intertidal restoration. We increasingly recognized how our actions in the world were shaped by the logics of temporality, categorization, and territorialization described above. For example, through listening to and standing with community partners, we identified what reciprocity means in practice and how to revitalize intertidal ecosystems and support numerous community-based restoration projects.[63] These projects are led by clammers and civic leaders and focus on activities like protecting and growing baby clams, finding ways to increase the diversity of fishing livelihoods, and addressing widespread water pollution. In projects focused on water pollution, participants often describe how water pollution is a justice issue, recognizing that mudflat closures impact clammers and Indigenous peoples more than affluent, also typically White, settlers and landowners.

Despite awareness of the complexities of water pollution, the sense of immediacy and urgency with this problem reinforces impulses to find technical solutions to mitigate sources of pollution, as opposed to attempting to address colonial land relations that perpetuate pollution.[64] Further, while the 1641 Body of Liberties described in the introduction challenges logics of territory by granting "shared access" to the intertidal, the racialized exclusions in who inhabits this territory create conditions through which intertidal access simultaneously reinforces land dispossession as a form of colonialism. These and a host of related examples remind us of the incommensurabilities of collaboration and worldmaking, especially when these efforts intend to support justice and when the ontological space of our engagement occurs within colonized coastal communities. In this context, we increasingly started to name this pattern as environmental imagination and found ourselves asking the above question about alternatives.

The tides have offered a partial and provisional answer to this question. As vital entities, coparticipants in these collaborative efforts, the tides taught us to notice the constraints of linear temporality as a key feature of environmental imagination. The temporality of tides is cyclical in a non-deterministic way, as tides manifest differently depending on the specific arrangements of weather, geography, season, the moon's position relative to the earth and relative to the sun, and so on. In its localized instantiations, tidal temporality is multiplicity. In this way, tides defy easy categorization. Defining tidal stage for a particular location never captures the ecological heterogeneity of the place, where the tides drain out at their own pace with endless variability. The tides, and by association the ocean too, push on the "colonial cartographic violence" that would fix the tides as a single thing and organize land and water into binaries, as Tiara Na'puti describes.[65] Finally, the agonistic negotiations of what intertidal access means for whom and how the tides challenge this too, especially when territory is defined based on the height of tides and the tides keep changing, has been a further lesson.

These initial lessons from the tides have been enriched by turning to the work of Black, Indigenous, and critical race scholars whose unique relationships with the tides and place-based histories have afforded insights about the value and necessity of tidal attunements described more fully elsewhere. Thinking with tides has helped produce attunements to the politics of our ecological engagements and contingent attempts at ethical worldmaking. Tidal imagination remains open to praxes that can challenge dominant logics of time, category, and territory by trusting in the emergent knowledge that comes from learning to listen as relation.

Thus, tidal imagination and knowledge coproduction share a commitment to listening. Showing up with an orientation to listening is what enabled the tides to start to do their work on us, to help us begin to feel the differences between environmental and tidal imaginations. Tiffany Lethabo King similarly characterizes the Black shoal as a methodology of listening that challenges "discovery narratives" and other colonial impulses in dominant science.[66] Instead, for her, "the shoal functions as a site that introduces new formations, alternative grammar and vocabularies, new analytical sites that reveal the ways that some aspects of Black and Indigenous life have always already been a site of co-constitution."[67] A shoaling, tidal orientation is an open mode of listening that is not about easy resolutions or coming to singular understandings. Instead, tidal imagination becomes a way of staying with the sites of "conceptual difficulty" to create "a process, formation, and space that exists beyond binary thinking."[68] This imagination promotes the necessary humility that comes with recognizing the racialized and colonial histories that inhere in commonplaces—like a beloved image of the earth, calls to unite humanity, and the environmental movement—to then, from there, reimagine how to create knowledge of the earth and other through incommensurate, yet still possibly ethical, relation.[69]

Microclimate 4 (Candice and Bridie)

The ideas flowing into and out of this space shoal around questions of what an ethical response might be from within vast, complex climates that reproduce the harmful logics and structures of racial capitalism and settler colonialism as atmospheric conditions permeating our lives. How can we better notice and understand, let alone disrupt, this matrix of power that we are complicit in reproducing, that circulates in the air and has seeped into our ground and foundations, and that has come to haunt our places and become naturalized texture woven into the minutiae of everyday life? What kinds of ethical orientations, capacities, and imaginaries are necessary for working toward a future that is otherwise?

Collaboration offers an initial response to these questions, a call that opens this book and brings us together as co-authors. Collaboration articulates a host of meanings that are appealing in times of isolation and precarity. Better together. Collaboration gets it done. Wholes, sums, parts, and so forth. Collaboration affirms the creative possibilities born of solidarity and shared sacrifice. Yet collaboration is also inescapably contaminated, as Anna Tsing puts it, "Collaboration means working across difference, which leads to contamination."[1] Collaboration toward things like justice, equity, or shared futures isn't about reaching consensus or "transcendent utopias." Instead, as our next chapter explores, "it is messy and endless and complicit." Collaboration, then, is a process of creation and destruction that threads through all we do. Collaboration as contamination means we're always already living in climates that we resist and seek to transform. We are shaped by them, unable to escape—breathing in the ash of wildfire, viruses, images of police brutality, unintended intensifications of inequity—and in that same breath taking in that which sustains life in this form, contributing to its perpetuation where sustenance and violence ravel relations. Environmental movements and academic institutions, considered in our chapters flanking this writing, are made through such relations and practices, as are all other social collectives and communities. The daily habits of where to focus individual and collective creative, destructive energies depend on ecosystems and histories that nourish some ways of flourishing while neglecting, diminishing, and erasing others. Being honest and humble about the conditions that make our collective lives possible, including our complicities within such mess, may orient us differently, obliquely, to collaborate otherwise: a reaching, "directional politics" to change the climates we find ourselves in, enacting climate change as contingent, ambivalent, and ethical praxis.

Thinking across ethics and praxis, then, offers a second response, especially when these ethics and practices are shaped by the oblique angle of tides, with their power to erode even the most hardened coast and shape daily habits. The coastal mudflats of Maine offer up a tidal ethics, a praxis that reimagines and works toward more ethical relations. Explored in the preceding chapter, tidal ethics are committed not only to "challeng[ing] dominant logics of time, category, and territory" but also to practices of collaborative worldbuilding that requires "trusting in the emergent knowledge that comes from learning to listen as relation." The next chapter explores an ethical praxis formed on the poisoned grounds of universities, one that aims to notice and to discover (or create) opportunities for disrupting the structures of racism and colonialism that are reproduced everywhere in institutional life. This ethical praxis imagines change through tinkering and ongoing small, mundane actions of people collaborating over time to recontour institutional life.

In our coming together here, there's also a common insistence that listening practices are necessary for the critical, ethical praxes we are imagining. Listening is a form paying attention to how the violent histories of racism and colonialism (and otherwise) shaped and persist within climates. Through listening, we also recognize that harm is reproduced not only through overtly racist practices but also through everyday interactions and institutional habits, and perhaps, most insidiously and painfully, through our acts of care and social justice (for example, in our sincere calls to unite humanity in the environmental movement or in well-intended literacy education aimed to increase access for historically marginalized students). Such listening requires humility. But this humility, which comes from understanding our own complicity and the contingencies of our labors, also opens us toward more relational and collaborative imaginations and helps us recognize the significance of modest ways of working toward alternative ways of being.

Tidal orientations point us toward capacities of listening and collaborating that are "not about easy resolutions or coming to singular understandings," but that model ways to stay within the complexities, complicity, contradictions, and multiplicities as we labor to transform. Tidal orientations also help us understand the importance of paying attention to the body. Listening for embodied knowledges, sensations, and affective energies helps us feel around for the fissures in the totalizing orders and sense places of incommensurability, jarring rhythms, and opportunities for intervention and for ethical, relational worldbuilding. Such listening practices are also important in the stories coming next. Listening for how institutions move (or not) is about noticing "all the mundane ways an institution is kicking itself back into place," a feeling registered when shoulders, necks, eyes begin to ache or twitch from the repetitious encounters with institutional resistance of many kinds. Sometimes these movements trace a subtle tick of difference, a rhythmic deviation as an institution kicks back into place at a slightly skewed angle, an oblique turn in its habitual trajectory.

In this way, one also listens for opportunities to intervene and recontour—for thinking with—the climatological surrounds that manifest in everyday life. Listening becomes an ethic that orients toward understanding, action, possible disruptions, change, however gradual or imperceptible. Listening with the body from places of complicity can also open space for empathy within critique. Perhaps because climate perspectives help us recognize that all beings are influenced by the shaping forces of their climate. A wondering about empathy streams into the next chapter as we shift to explorations of the coming together of racism, colonialism, and violence in exclusionary language ideologies in literacy education.

In our coming together here, we consider how to transform places and communities while staying with the incommensurabilities and complicities. Is it possible to see as kin someone holding an idea you find vile? Is it possible to empathize through critique and transformation without reinscribing forms of hate, exclusion, vilification? Is it possible to hold onto the multiplicities and still be oriented toward more ethical relations? We hope so: it's a hope we hold dear through tidal imaginations, institutional transformations, and tinkering otherwise.

5 Institutional Climate Changing

(Candice)

Climate, n., from Greek, *klima*, inclination, lean, the general weather patterns and conditions within a place.

Institutional climate, n., the leanings, commonplace opinions, ideologies, values, languages, and stories that circulate, inclining us this way or that; the habituated ways of knowing, doing, feeling, and being together that nourish or diminish lifeways within an institution or other social place.[1]

Our universities were planted on poisoned ground. We work on stolen land. Our buildings are placed on the ancestral lands of Indigenous peoples, built over graves and sacred places, perhaps with wealth accumulated through the Atlantic slave trade or through the labors of enslaved human beings or by people of color, immigrants, the poor, and others, who had no access to the places they helped build. Structures of racism; logics of settler colonialism; linguistic, cultural, and class-based discrimination; and other inequities endure in our places of learning, despite generational labors to reshape access and inclusion in ways that acknowledge these exclusionary, violent histories and that create the conditions not for acclimation to these hostile climates, but for transforming them into places more of us belong and thrive just as we arrive. Climates are changing, yet the labor continues. This: a spare history of the problems of access, the persistence of racist climates within US universities, and the ongoing work of transformation.[2]

Literacy and literacy education has long performed racial violence and exclusionary functions in the United States under the guise of objectivity, neutrality, and even the public good. This is evident everywhere—from the cultural and linguistic erasures experienced by Indigenous communities in boarding schools to the voter literacy tests designed to disenfranchise Black and Brown people, immigrants, and the poor and to the ways that racist language ideologies and colonial epistemologies entangle in the teaching of White cultural and linguistic norms globally.[3] Catherine Prendergast identifies literacy as an enduring "site of the struggle for racial justice" because "throughout American history, literacy has been managed" to protect "White property interests, White privilege, [and] . . . America as a White nation."[4] As Suhanthie Motha emphasizes, "Racism has become so naturalized within the project of teaching English that its presence is no longer noticeable."[5] Under the "guise of correct grammar . . . and unquestioned language hierarchies, we have created an ideal climate for Bonilla-Silva's . . . 'racism without racists,'" where teachers, often unknowingly, "sustain racism" by centering language practices "that are most richly rewarded" (namely, Standard U.S. Academic English).[6] Asao Inoue calls out the racist language ideologies rampant in the university and our broader culture that "clearly privileges middle class white students" and inform the foundation of most U.S. writing programs.[7] Despite intentions, Inoue reminds us, if we're a "part of larger institutions like education, like the discipline of composition studies, or the teaching of writing in college," then we're participating in reproducing racist structures.[8] While BIPOC and other marginalized students have benefitted in various ways from learning the dominant linguistic norms taught in our schools, as many have pointed out, this often comes at great cost in terms of experiencing trauma, loss of home languages, and cultural alienation.[9]

This chapter leans into these legacies of racist climates within the university as they manifest around language ideology and the teaching of writing and rhetoric in a large primarily first-year writing program at the University of Washington–Seattle (called the Program in Writing and Rhetoric [PWR]).[10] Here you will find stories about the ongoing, collaborative work we have engaged in to transform these climates toward more equity-oriented, inclusive, and just inclinations within our writing program.[11] The tales shared here occurred in the time that I directed the program from 2014 to 2021, but this ongoing work builds and relies on the leadership of my colleagues, Anis Bawarshi, who directed PWR right before me, and Stephanie Kerschbaum, who now holds the baton with grace and intention. This period was and continues to be a politically tumultuous time, signaled by the intensification of the conservative right, White supremacy, xenophobic ideologies, racial violence, and economic inequity, as well as by the momentum of the collective movements of Black Lives Matters and for racial, economic, and environmental justice. Referring to writing program administration as "always and already race work," Staci M. Perryman-Clark and Collin Craig ask those of us charged with leading or teaching in writing programs to consider how our work is either actively resisting or "function[ing] as citations of larger political projects to disenfranchise historically vulnerable and marginalized groups."[12]

In response to these political urgencies, this essay asks how we might collectively transform racist institutional climates (within the university and beyond) toward more just, equity-oriented, accessible, and antiracist learning environments. Such is the work that many writing programs, universities, and other public institutions are currently engaged in across the United States. While with reenergized momentum in this time, this work is only possible because of the ongoing efforts of so many over years, decades, entire lifetimes—people and communities who resisted and said "no" at great cost or who labored with their precious life energies to lay the foundations that make any climate change we might hope to accomplish possible.[JL] In thinking of our modest collaborative work, many stories come to mind of things we accomplished (or failed to) within the humble context of our writing program, the labors of incremental change, revision to this policy or that, another workshop or conversation, curriculum to revise, biases to reflect on, things to unlearn, spaces for listening to create, mistakes made, complicity, half measures, momentum, exhaustion, so many things. I'm reminded of what Sara Ahmed said of antiracist work in institutional spaces: "No wonder that antiracist work can feel like banging your head against the brick wall. The wall keeps its place, so it is you who gets sore."[13]

Yes.

But I also think of Christina Sharpe's wisdom. In one vein, her work helps us better grasp the impossibly long life and all-enveloping nature of anti-Black racism "in the wake" of slavery. "Antiblackness is pervasive as climate," she says, the "atmospheric condition of time and place."[14] And, in another vein,

---

JL Returning to your opening definition, the metaphor/physicality of "leaning" or being "leaned upon" captures so much of what it's like to inhabit a climate full of micro and macroaggressions, and it gets at how resistance is not just a mere reversal. Rather, you have to push against the lean before it is even possible to reroute energy in a different direction.

when speaking of "wake work," she gifts us a praxis for imagining "new ways to live in [this] wake," a "mode of inhabiting *and* rupturing" and for imagining "otherwise from what we know *now*."[15] Contrasting the ongoing violence Black people have experienced for simply "surviving" or "for daring to claim or make spaces of something like freedom," Sharpe offers wake work as a way to "reimagine and transform spaces for and practices of an ethics of care (as in repair, maintenance, attention), an ethics of seeing, and of *being* in the wake of consciousness."[16]

Thinking about the impossibly small actions in a writing program as a form of wake work seems a humbling transition and one that might be met with criticism. One thread woven throughout this climate essay is the question of the role that literacy education and language ideologies, practices, and policies play in maintaining the long history of structural racism that persists in our university climates. Speaking to these concerns, April Baker-Bell argues that

> the way Black language is devalued in schools reflects how Black lives are devalued in the world . . . [and] the anti-Black linguistic racism that is used to diminish Black Language and Black students in classrooms is not separate from the rampant and deliberate anti-Black racism and violence inflicted upon Black people in society.[17]

Another thread woven throughout this climate chapter, then, is the question of what those of us implicated in this violent history can possibly do to resist and transform the pervasive racism and other structural exclusions in our own institutional places.[CI] With these threads in mind, what follows are stories of modest actions, meditations on ethical orientations, and explorations of the possibilities and ambivalence of equity-oriented institutional climate change work.

CI Candice is being modest about this being a "thread." For many of us, and no one more than Candice, it's a veritable commitment. "What can be done" seems to me to be both the question and answer of this essay, which is another one of those recursions that climates also exhibit. In the same way that being in a climate contributes to creating the climate that we then go on being in, asking "What can be done?" itself contributes to establishing the importance of the answer.

On Racist Climates and Climate Change in Institutional Life

Racist climates, n., All-encompassing conditions that perpetuate systemic exclusion, inequity, erasure, and violence based on race and other intersectional identities; climates that reproduce structural inclinations and atmospheric particulates that kill, thwart, promote failure to thrive, and impede survival based on race.

Climate change, v., Laboring to change racist and otherwise toxic climates (political, institutional, environmental, etc.); using precious life energies to change the climatological leanings and the conditions of survival.

Ex.: *We've been climate changing for centuries, yet racial violence remains pervasive in our institutions.*

Climate changing is hard. But you already knew that.

Let's return to those walls that Ahmed writes about—the ones you hit when doing antiracist and anticolonial climate changing work. Every which way you turn: Another wall. Blamo! A wall "gives physical form," as Ahmed puts it, to "'institutional inertia,' the lack of an institutional will to change."[18] As she writes, the "habits of the institutions are not revealed unless you come up against them," otherwise institutional habits keep rolling along because "when something becomes a habit . . . it saves trouble and energy . . . it does not have to command your attention."[19] The collective habituation of actions is how an "institution takes shape as an effect of what has become natural."[20] This accumulation of naturalized habits is what we might call the climate.[BM] As the above definitions imply, racist climates are characterized by the conditions and patterns of collective life that perpetuate intersectional racism. Climate changing includes resistance, action, and worldbuilding labors required to imagine a world otherwise, as well as to recontour everyday life toward more

BM We intentionally define climate in many different ways across the volume, which is an impulse I appreciate and agree with. But I find your definition here of climate as naturalized habits to be particularly sticky and helpful, especially for thinking about how something like a visual image feeds into climates, as habits unnoticed as habitual.

equitable futures. Climate changing, in this sense, "creates trouble" because it disrupts long-standing norms and ways of life.[21]

La paperson's work in *A Third University Is Possible* resonates with Ahmed's thinking about institutions as always already exclusionary and as also providing the opportunities for transformation from within.[22] La paperson writes of universities as "assemblages" and "bits of machinery" that can be messed with by "subversive beings" that can hack, "wreck, scavenge, retool, and reassemble the colonizing university into decolonializing contraptions."[23] La paperson calls these subversive beings "scyborg—composed of s [for system/structural] + cyborg . . . to name the structural agency of persons who have picked up colonial technologies and reassembled them to decolonizing purposes."[24] "Scyborgs are system-interference and system-witchcraft, the ghost in the machine."[25] Like Ahmed, la paperson doesn't see institutional climate-changing work as necessarily revolutionary, though it can lead to more just futures. Moreover, precisely because the scyborg is of the machinery, involving those with power in the university, and because equity problems are reproduced pervasively throughout the climate, the possibilities and agency for system tinkering reside everywhere. While always complicit and partial, tinkering does something and that's not nothing.[JL,BM]

While probably surprising to no one here,[JA] it turns out the core mechanisms of writing programs and universities are coalescing everywhere to perpetuate a

---

JL Yes.

BM I hope it's ok that I'm tagging onto JL's note here too, because I had the same response: Yes. And I think the way you're describing this, as tinkering, as actions that do something while attempting to also sense how these actions are always already complicit, is so much in the spirit of what la paperson call us as academics to do: to ground our praxes in decolonial love and turn our tinkering to decolonizing the academic machines we inhabit and that we can also change through our choices, contingent as they may be.

JA Perhaps it is surprising if the reader has never worked in a writing program at a public research university. I don't mean to undercut other site-work because white, patriarchal, straight, ableist oppression circulates in department structures, student life and housing, the managerial class of campuses, sports and so forth, which is why I think la paperson implicates us all in the settler apparatus of the university. If you work for a university you are implicated; if you live in the metropole, you are implicated. But my comment is more specific: there is a level of pain and frustration in doing this kind of elevated antiracist, difference-inclusive work within a writing program (most often housed beneath an English department) and askew from knowledge "innovation" and so forth. Antiracist work from the basement. This

climate of racism, ableism, coloniality, and inequity. Indeed, anyone who has engaged in antiracist writing program work already knows that racism and other structural exclusions lurk in every program blurb, rubric, manual, interaction, in all the criteria, reports, assignment prompts, in every bullet of every learning outcome, in every space created for certain bodies and not others. Climate changing, in the antiracist institutional sense, is like hitting walls; or maybe more like being held in place by a million tiny kicks;[26] or like shedding skins that become so many abandoned forms of ideas, structures, practices, habits, no longer of our bodies exactly, yet sticking around by a thread here, a thread there, refusing to let go, unshakeable forms being dragged about, rattling skin bags, empty yet retaining the forms of violence we tried to shed. Ghostly presences that can pop up like zombies in a haunted house, always threatening to revivify.

Okay, perhaps all I'm saying is that institutional transformation is an exercise in extreme humility and wherewithal.

But it also feels, at times, like creative, collaborative praxes to actively notice all the mundane ways an institution is kicking itself back into place (and then to mess with those things, poking away with your broomstick). Personally, it's also very humbling work, especially when I come to discover that it is me (or our writing program) doing the active kicking back—doing the thing, saying the thing, teaching the thing—and suddenly recognizing myself as an agent of institutional habits that perpetuate harm, often the very ones I'm trying to resist, and sometimes through actions performed in the name of social justice. As Max Liboiron reminds us, "our work is already compromised" and "caught up in the contradictions, injustices, and structures that already exist, that we have already identified as violent and in need of change."[27] Raewyn Connell, on whom Liboiron draws, makes the point more sharply: any attempt to reform "social institutions and systems from scratch, in a blank space" is the "colonizer's dream."[28] Michelle Murphy also rejects the possibilities of non-complicit transformational work, advocating for a "commitment to act now . . . to invent, revive, and sustain decolonial possibilities and persistences right now as we are, forged in non-innocence."[29]

Ahmed's willful politics offers another touchstone for ethical-oriented action in step with these concerns. Willfulness is a "political art, a practical

---

is where I've worked for so long now—which is why I am in awe of the curriculum, culture, and community on display here. Thank you, Candice.

craft that is acquired through involvement in a political struggle, whether that struggle is a struggle to exist or to transform an existence."[30] Ahmed highlights all the willful micro actions and shifts (e.g., climate-changing labor) to environment, bodies, language, ideas, and relationships that are necessary to recondition the world. Specifically writing about U.S. slavery and anti-Black racism, and the infinite small actions required to conceive an alternate world, let alone remake it, she writes that a "collective will can only be realized through individuals who are willing to push back in order not to be pushed into obedience."[31] Willfulness politics is also about the climate-changing labor people perform to persist into the future and, therefore, can become "what travels, as a relation to others, those who come before, those who come after. . . . A political action can be what is performed to stop a chain from breaking. The individual capacity not only to say no but to repeat the no in what bodies do not do could be described as a willful gift."[32]

I resonate with this idea of a praxis that keeps a chain going: you pick something up and continue as best you can in order to pass it on to others. I like how relational this image of working is, too—a reaching back, a reaching within and among, a persistent reaching toward so that others might be able to better reach or to say no or simply keep working or surviving into the future.[JL] It is a kind of directional politics, an inventive capacity of resistance, action, and care within all-encompassing climates, an ethical orientation, a grounded, entangled sort of politics in which one reaches with others from thoroughly complicit places toward a world not yet present, nor even fully imaginable, let alone realizable anytime soon, if ever. This praxis helps me hold onto the value of small, place-based, mundane actions as mattering. But this is no praxis of hope born from transcendent utopia or a grand revolution. It is messy and endless and complicit.[NS]

JL To return to the point about "tinkering" as doing something, not nothing, it is possible to reach while you lean, even if it strains the back.

NS I really like these turns on the compromised and complicit nature of refusal and contestation, the necessary non-innocence of it. And I really like how you are taking us through the problem in relation to literacy. To me this is not dispiriting because it means letting go of that impossible dream of ex nihilo invention, free of the ethical compromises that somehow, nonetheless, brought one to seek such invention. It raises for me another question which I have found that most folks consider unwelcome if asked. How can we inhabit violence ethically? Education is grounded in violence as you have set forth so effectively. To me the question is how to understand education as ethical violence—not as colonizing, not "for

Keeping hold of the above threads, I'd like to turn our attention back to the context of climate changing within writing programs. Acknowledging the long-standing violence of literacy education, our complicity in reproducing these harms, and our commitment to ongoing transformational work is the starting place for antiracist, anticolonial work in our writing program, which is reflected in the following excerpt from our antiracist program statement:

> We acknowledge that literacy education and language policies in the U.S. are built on a foundation of racial capitalism, White supremacy, and settler colonialism that persists and has delegitimized and often penalized the language practices, experiences, and knowledges of minoritized and historically underrepresented peoples. . . . We seek to transform this ongoing systemic inequity and discrimination by developing writing curriculum, assessment practices, teacher development programs, and language policies that recognize linguistic and other differences as the norm of communication.[33]

This statement[JA] was collaboratively written by Sumyat Thu, Emily George, and me in early 2018 and has since been revised with input from many.[34] It was crafted in response to the political urgencies mentioned earlier and most directly in response to the aftermath and charged political climate that erupted in the wake of Trump's presidency. Immediately after the 2017 presidential inauguration, our campus, like so many others, experienced eruptions of racial violence, including a nonfatal shooting at a campus protest, hate speech, instructor doxing, and antagonism in classrooms. Drawing on antiracist, anticolonial, translingual, and intersectional feminist approaches to language and writing pedagogy, this statement was intended both as a support for instructors already doing antiracist, equity work in the classroom, and as a generative launching pad for collectively defining our commitments and for guiding our reflection, action, policies, practices, and

---

your own good," but action which cannot help but arise from, respond to, and risk violence, yet done ethically? The seeming impossibility of asking that, of considering violence in its ethicality, reflects the dream of innocence calling us to sleep, I think.

JA And this statement will be read, discussed, rewritten, and posted in the Program for Writing and Rhetoric at University of Colorado Boulder, where I work—if we have the fortitude. Thanks for the precedent.

pedagogies. It was something we could keep revising and critiquing—a set of commitments to which we could hold ourselves accountable.[35]

Driven by the commitments expressed in the statement, we began interrogating our program, making plans for intervention, and building material supports for sustaining this work. Below is small list of some climate-changing work we engaged in:[CI]

> We began by taking stock and creating an action plan: we reflected on program documents, vocabularies, manuals, curriculum, professional development, etc., to ask how these genres and practices (and we) as agents of institutional habits might be perpetuating racism or otherwise inequitable structures. In addition to accountability, this ongoing reflection was intended to help us continually discover concrete places for resistance and transformation.
>
> Emerging from the above, we prioritized building more capacity in our teaching community for doing equity-oriented work by revamping our teacher orientations and pedagogy course to center issues of equity, inclusion, and accessibility and to make antiracist praxes the core organizing frame for everything we do as writing teachers.
>
> Finally, we worked on creating material supports and processes to support our equity work in ways that were more ecological, material, structural (climatological), and thus more likely to continue as habits. We also tried to find ways to acknowledge the small changes that accumulated and build opportunities for continual reflection and action. For example, to build capacity and energy for this work, we created a repository of sample teaching materials of equity-oriented work, launched a pedagogy workshop series and internal grants to support equity-oriented teaching practices, supported an endorsement in critical pedagogy, among other things.

CI Pausing here, though it could well have happened elsewhere, to say how much I love the nods to "climate change" in its traditional register, which usually means something to do with a heating earth system and human excesses. But here, as a participle—"climate-changing"—there's far more happening. There's movement, an ongoing commitment to effectuating that change by design, instead of as a byproduct of indifference or ignorance. And that does, of course, take work. It can't happen passively because, unlike petro-capitalism, it's not already baked in or systemic.

We even discussed concerns that the antiracist institutional transformation work we were doing could become new habits that need tending to, lose transformative vitality, or perform unintended harm. To this point, it's been important to recognize how equity-oriented work keeps folding in on itself, producing new problems, which is something to just sit with, I suppose, as one deepens into accepting the ongoing/complicit/always-tainted nature of institutional transformation.

As one example of this, I share a story, which I could title, *How My Justice Work Created New Walls and Harms: The Endless Labor.* This vignette has to do with work I now do as faculty coordinator for the UW in High School (UWHS), a dual enrollment program that supports around fifty high-school teachers throughout Washington state who offer our writing courses for college credit. We've been working with teachers on translating our antiracist commitments from campus to their secondary education contexts and navigating the challenges of doing so. During a conversation with the group, one teacher offered that she couldn't help feeling that we were all exacerbating structural inequity, despite her and our commitments to equitable teaching. Why do you feel that way, I asked? She explained that, for complex reasons, the students who end up taking the UWHS courses at her school are majority White, middle class, and already successful users of Standardized English. We all sighed. Yes. While I know this is not the case for all of our schools, I saw many nods and others confirmed a similar dynamic in their classrooms. I asked, What do you think we can do? And . . . pop: another trapdoor of whoop-ass opens to an abyss of labor awaiting us, just now peeking into view on the distant horizon, the tip of another iceberg, another wall, or opportunity, you could say. *Should I close that door? Look away? Would you? Maybe no one else took much notice. Not in my job description? I'm tired.* This kind of climate-changing labor takes attention enough just to recognize the problem and then lots of labor beyond to even begin addressing. But you already knew that.

Part of the concern here is a wondering about how to persist in climate-changing labors, while accepting the active kicking back into place happening every which you turn, and while recalling that any changes you make simply sow the blooms of next season's problems or produce new forms of violence. The next section picks up this thread and deepens into the ambivalence of equity-oriented climate-change work within racist institutions.

On Obliquity and the Ambivalence of Equity

> obliquity, n., From Latin, *oblīquitās*, to be slantwise.
>
> Earth's slant, the angle between Earth's equatorial and orbital planes, the axial tilt determining where the sun shines.
>
> Slantwise existences marked as deviant from the dominant accepted ways of knowing and being that sustain and are sustained by a climate.
>
> Willful deviation from the norm, possibly to ensure existence or to bring about a new existence; labors, often hidden in shadows or on the periphery, intended to change the conditions of survival, possibly in the name of justice or for devious aims or both simultaneously.

Obliquity, as it relates to environmental climate, is Earth's tilt as we travel around the sun and the reason we have seasons. Obliquity shifts cyclically over long periods of time, affecting climate. It also carries social meanings, as you can see in my above definitions, that help reveal the ambivalence of climate changing. All political climates have strong slants—call these biases, inclinations, ideologies, norms, ways, means—that are reproduced within a climate and shape lifeways. Deviating from these slants (willful or not, whether to carve out alternative existences or simply to exist) is to risk being dismissed, marginalized, feared, hated, censured, or worse. Thus, willful action often happens on the fringe or starts there, which gets at the sense of obliquity as opaque, hidden, emergent, sneaky, suspect sorts of labor and existences. The other relevant part of obliquity as it pertains to political climate changing is that there is generally a deep ambivalence around justice questions. Whose slant orients toward justice and equity (and how are such things defined and by whom, anyways)? How do we proceed when incommensurable inclinations clash or the conditions of survival compete? What do we do when shifting where the sun shines means that while it becomes more livable and abundant over here, for us, it becomes less habitable over there, for them?

Despite the intractable political tensions embedded in these questions, there is also the possibility of transformation implied. For if Earth's obliquity can wobble, shifting the planet's climate, then it must also be possible to transform the racist climates in our public institutions. While the kind of antiracist institutional work I've been talking about here is never finished, it has been heartening to see some momentum within social justice and equity movements nationwide and within the small ecology of our writing program. In this moment, we can point to transformative antiracist practices everywhere in universities, professional organizations, and broadly in the public—such as changes to hiring practices that support faculty of color, increasing representation of Black and Brown people in the media and in leadership positions, and so on. Every which way we turn we can see evidence of antiracist climate changing. Both in terms of concrete changes we can point to and in ways more atmospheric, it feels like the slant and inclinations of our climates are changing in the time I am writing this.

Of course, not everyone is celebrating these changes. For some, this is alienating, evokes fear, feels violent, disrupts lifeways. Some have met these tilts with backlash and work to climate change in opposing political directions (e.g., in this moment, anti-LGBTQIAA+ legislation, the undoing of *Roe v. Wade*, legislation dictating how race is taught in schools, etc.). Here, the ambivalence of obliquity and equity-work helps us better understand why antiracist climate changing is met with such fierce resistance and why this opposing resistance is often also performed in the name of justice, equity, freedom, and/or the public good. This ambivalence of equity work and its inherent political clashes scale up and down in terms of scope and violence, perhaps, because tilting the leans of a climate means shifting where, how, and for whom sunshine, privilege, and comfort is experienced, as well as whose ideas, cultures, and lifeways are sustained by the climate (or not); in more extreme terms, this can mean changing the very conditions of life and survival. (Re)contouring the climate to better nourish some, presumes less nourishment for others.

The insights on trophic rhetoric from our earlier "Inclement Weather" chapter seem relevant here. As Nate discusses, trophic relations are the complex, dynamic ways that all those alive (or who once were) within an ecology are interconnected and codependent, exchanging finite nutrients and energy, either feeding on or nourishing one another. Nate draws on Caroline Gottschalk Druschke's work, the latter arguing that trophic relations, in the ecological sense, draw our attention to the "flows of energy transferred in forms of relationality between beings through hunting and eating and fucking and decomposing," which shifts our understanding of rhetoric to "a connective verb composed of physical, palpable, symbolic, affective, and chemical relations."[36] Nate describes trophism as "nested interdependencies" and "relational worldmaking," emphasizing that the "logic of nutrition that informs trophism deflects from the fact that trophism is inherently violent (what feeds on what and under what conditions), which raises the question of how forms of violence sustain and unravel ecologies."

This question of how violence is sustained is also centered in John's chapter to come, which explores the concept of the ossuary as the biological commons indexing violence as an inescapable logic and as indelible within a climate. The idea that violence can somehow be overcome is abandoned in this thinking, though its forms or the balance of who or what experiences (or deals out) more or less of it may transform. If we find collective ways to stem or redirect violence, at least here and now, this is not permanent or transcendently virtuous because whatever it is we have done to keep violence at bay (or move it elsewhere) creates its own problems for someone or thing else. Our reading group's conversations have taught me much about violence and what it means to act in the face of its intractability. While my disposition keeps me circling back to pragmatic questions of hope and transformation, those things mean something very different now. Perhaps you'll wonder with me: What happens to questions of justice and action when you accept violence and its remainders as the constant? What does it mean to hope, act, and persist with the knowledge that your sense of justice and your equity-oriented actions perform violence? Or that no matter how you try to mitigate harm, your work will reveal itself as poison to someone or thing or in some place? Given this, how do we conceive of ethical orientations? What could possibly motor our aims and desire for social justice?[NS]

NS These questions! Yes. I feel like you have captured the burdens so crisply. You asked me in my chapter how I face oblivion given the inevitability of harming others, but I think your questions are that answer. We are thrown into such questions by what we cannot surpass and so charged to live still. I think, for example, of all the harm done to me by people who genuinely loved me, in their love for me, and I cannot say that because violence was done that I am broken. I am lucky in that, but my point is not about luck. Violence is never encountered qua violence for me. Living with violence is not defeat. Living with it differently, being better, accepting that we will hurt others (and in some cases need to) should not be, for me, an invitation to be equivocal or resigned about injustice and its many horrors. The violence I have experienced (and done) is not uniform with every other violence. After all, violence is a quality of relation, not a species of acts that can be replaced.

With these ideas on trophism, obliquity, and the ambivalence of equity work in tow, let's turn our attention back, once more, to climate changing within education institutions. The increased attention to antiracist work that many educators are currently taking up across K–12 and university contexts in the United States is being met in this moment with a palpable backlash. One recent example is Florida State's legislation, An Act Relating to Individual Freedom, colloquially known as the Stop W.O.K.E Act (Stop the Wrongs to Our Kids and Employees Act), which went into effect on July 1, 2022, as a response to the kind of critical race theory–informed pedagogies and practices our writing program and many others are working toward. Florida Governor Ron DeSantis refers to this act as "the strongest legislation of its kind in the nation" that "will take on both corporate wokeness and Critical Race Theory."[37] The bill attempts to legislate how race is discussed in school and workplaces. For example, it deems "unlawful" to subject "any individual . . . to training, instruction, or any other required activity that . . . inculcates, or compels such individual to believe any of the following:

> An individual's moral character or status as either privileged or oppressed is necessarily determined by his or her race, color, sex, or national origin; . . .
>
> An individual . . . bears responsibility for, or should be discriminated against . . . because of, actions committed in the past by other members of the same race, color, sex, or national origin . . . [or] must feel guilt, anguish, or other forms of psychological distress because of actions, in which the individual played no part, committed in the past by other members of the same race, color, sex, or national origin. . . .
>
> Such virtues as merit, excellence, hard work, fairness, neutrality, objectivity, and racial colorblindness are racist or sexist, or were created by members of a particular race, color, sex, or national origin to oppress members of another race, color, sex, or national origin.[38]

While there's much one could unpack here, I want to highlight the competing slants in this legislation in relation to our antiracist writing program work. If antiracist institutional work begins with the premise that institutions perpetuate and sustain racist, White supremacist, colonial, and other harmful logics and with a call for decentering and resisting these logics (including the decentering of White language norms as the tacitly superior standard in literacy education), then with its opposing political slants, what is at the heart of this law? At minimum, the law resists the core premise of institutional racism itself—which is that institutions sustain White privilege and exclusionary structures that we all participate in reproducing. It also makes it illegal in educational contexts to "espouse" the idea that we all bear responsibility for transforming our racist institutions. It seems important to acknowledge how this backlash may also be driven by trophic logics: fear that antiracist climate change—which focuses on dismantling systemic harm and generational privilege, decentering White cultural and linguistic norms, etc.—threatens to wobble the climatological tilts and distribution of nourishment, therefore, changing the conditions of life and survival.

In our writing program, aspects of our "Statement on Antiracist Writing Pedagogy and Program Praxis"—along with our program pedagogies, professional development, and curriculum—would be deemed illegal according to this Florida legislation. For example, our program is guided by tenets of critical race theory and the premise that our institutions were built on and continue to reproduce structures of racism that we all participate in upholding and are all responsible for transforming; that racial colorblindness contributes to the reproduction of racist institutional structures; and that public institutions generally sustain racist structural inequities by centering dominant White linguistic and culture norms (as though they are transparently universal or value neutral discourses/ways), among many other ways. Below are a few excerpts from our statement that reflect these ideas:

> We acknowledge that *literacy education and language policies in the U.S. are built on a foundation of racial capitalism, White supremacy, and settler colonialism that persists and has delegitimized and often penalized the language practices, experiences, and knowledges of minoritized and historically underrepresented peoples.* We therefore reject Eurocentric assumptions about the written word as a superior form of literacy and define composition and literacy in our program ecology as multi- or trans-modal, translingual, anti-colonial, and culturally affirming communication practices.
>
> Rather than being simply a matter of individual biases or prejudices, *we understand that various forms of oppression are pervasive, intersectional, and built into our educational, economic, and political systems.* Racism, sexism, oppression

> of gender nonbinary and queer people, ableism, and oppression on the basis of language and citizenship all work in intertwined ways to reproduce the conditions of racial capitalism and colonialism. *These systemic oppressions are ongoing problems that concern all of us, that we all participate in perpetuating even unconsciously and unintentionally,* and that require us to understand the important differences between intent and impact.
>
> We seek to support our students and instructors through active antiracist and equity-focused pedagogies and program praxis that: . . . integrate language justice work as part of writing courses in which we examine how *systemic racism is often encoded in practices that uphold "academic language" or "Standard English"*; . . . encourage students to think about the *social impact of their writing and the social groups and communities they are accountable to as part of audience awareness.*[39]

Another relevant thread in our program statement resists the idea that writing and rhetoric instruction can ever be objective, neutral, or apolitical, as is called for in the Florida legislation. In fact, we are explicit that so-called "objectivity" and "neutrality," in fact, can (i.e., does) serve to reproduce, naturalize, justify, and obscure racist structures. We also push students to think about how their positionality affects their writing for diverse communities and to be accountable for these impacts as part of audience awareness, a learning goal that is assessed.

Clearly, the slants of our statement and that of Florida's Act Relating to Individual Freedom are at odds. In evoking the topos of freedom and turning Civil Rights rhetorics against those working to address long-standing racism within educational and other public institutions, we see in this Florida legislation another example of the ambivalent core of democracy. More could be said on this, but I want to focus on putting these clashing understandings of education and equity—these contested climatological slants—in conversation with recent experiences we have had in translating our antiracism writing praxis commitments within high-school contexts.

Housed within Seattle's liberal political bubble, our university writing program has not experienced much backlash for its antiracist work. However, high-school teachers in more conservative communities throughout our state who teach our courses through the UWHS program have concerns. There are very different challenges and resistances of doing antiracist work on a politically liberal campus where such work is both supported and expected these days versus doing it in diverse secondary education contexts across the political spectrum where minors and parents are also involved. We work with teachers in high schools with student populations that are majority White and middle class and teachers in Tribal and Title I schools where students are multiply marginalized and will be first-generation college students; we work with schools in liberal Seattle and surrounds and schools over the mountains in rural and farming communities with politically conservative views. Each context comes with its own unique opportunities and challenges.

In a recent professional development on antiracist writing pedagogy, several teachers expressed concerns about how to navigate our program's antiracist praxes in their politically conversative communities where there was resistance that parallels sentiments expressed in the Florida legislation. A couple of teachers said that even using the phrase "antiracist" could lead to trouble, and a couple shared narratives of parent or school complaints they had experienced for their critical views and pedagogies. Various questions were raised and discussed: *Is it possible to engage in antiracist praxis within teaching contexts that demand apolitical classrooms? How do you do antiracist work in hostile climates? Or, in places where your students are primarily White and/or politically conservative?* Honestly, I say, I do not know. We talked about how there's no template for antiracist teaching and institutional change; it's situational and placed-based, there's no one way, we have to work with our own and our students' positionalities and relationships to power in mind. Here, the earlier conversation about directional politics remains helpful. Antiracist, equity-oriented institutional change is not about unrolling a plan or getting it right. That is not possible; it is ongoing work. We work collaboratively, reaching, recontouring, refusing, continuing chains of labor, guided by ethical orientations that may point toward certain commitments (such as those expressed in our antiracist praxes statement) but that are also dynamic, open to interpretation, and need to be activated in places.

Equity work also needs to be continually reflected on. I can think of many examples of so-called progress that left us wondering what new equity problems we were creating. For example, beginning in 2020, we supported a collaborative process to revise our program portfolio rubric to better align with our equity-oriented goals by centering students' linguistic and cultural resources; emphasizing learning in process, revision labor, and development over performance; resisting the privileging of static monolingual language norms; and foregrounding ethical communication practices. In doing so, we and instructors also questioned the extent to which potentially ableist notions of labor are wrapped up in our articulation of "revision labor" and how those ideas are tangled with racist structures, too.

Another story comes to mind when reflecting on our equity work back on campus, given the clashing slants in our political climate. This story gets at how to make sense of moments when entire worldviews collide and helps illuminate ways that institutional climate changing work involves structural change as much as emotioned, affective transformation of imaginations, bodies, and interiorities. In particular, what do you do when a student introduces logics of racism from their cultural context into the very classroom space that you are actively trying to make inhospitable to such logics? For example, let me share an anecdote from my time as writing director. One day, a first-year college student came to my office. He had apparently shared an opinion that was widely accepted in his cultural context but that others felt was racist and his instructor and classmates called him out. Every time he tried to explain himself, things got worse. Do you understand why people felt the way they did, I ask? He said that he does understand more about why people were upset. I encouraged the student to talk with his teacher just as he had with me, but he didn't feel comfortable. I tried to talk to the student about how I could understand his standpoint and also his classmates' and shared some thoughts on navigating the situation. He didn't want me to intervene. Just thought I should know. I offered to talk again but never heard back.[40] My experience with such situations has informed teacher training and mentoring work I've done around empathy and rhetorical listening and to push against traditional argument models that tend to support entrenchment of positions rather than supporting capacities for working across difference.

What are we to do with this situation? Was this a missed learning opportunity? Was this a moment in which a harmful ideology was jiggled free a bit or further entrenched? What kind of climate-change labor should we engage in here? I'm not sure. But this student's insistence that his view was so "normal" to him has stayed with me. My sense was that he felt ashamed and defensive all at once. It reminded me of Jennifer Trainor's powerful study of students in an all-White high school, which illuminated how critical pedagogies sometimes backfire and reinforce, rather than combat, racism. Students' beliefs about race are not stable, she argues, but in flux, and they are also tethered to deep emotions and affective responses.[41] If challenging structures of inequity is ideological work—it feels important to remember that we sometimes ask students to challenge deeply held worldviews that circulate like the air within their home places, which is to say that it might be helpful to perceive views as climatological, patterned, and hardwired-ish, yet dynamic and shiftable over time, long periods of time. Political slants aren't flimsy ideas; rather, they are tied to home, loved ones, language, identities, place, everyday life, to vastly complex climates. Yet, we so often treat (and teach) beliefs as things we can just persuade or shame people out of with sound arguments backed by evidence.

Personally speaking, my own investment in resisting structural inequity on behalf of students who have been historically marginalized sits in uncomfortable tension alongside my inclination to extend more understanding to the above student's experience. We know that rhetoric is deeply emplaced, climatological, tied to materiality, everyday ways of knowing and being—but it is easy to forget that when someone says something racist in your classroom or elsewhere. It is common enough for teachers engaged in antiracist work to understandably say that they just cannot expend energy caring for the journey of students who they perceive to hold privilege. I certainly feel this sometimes, too. Still, this student's experience: it stays with me. I pause and sit in this place of clashing tilts and ambivalence. How might a rhetorical climate perspective inform our understanding and response here? Changing deeply held worldviews requires hard, ongoing, system-wide, emotioned labor—climate-changing labor—to shift the obliquity of a place. Climate is all-encompassing. We can't put our finger on all the ways it flows through us. There is no clear boundary between me (or you or us or them) and the shape of the climate, those powerful shaping slants, those leans and tilts that incline us. Shouldn't this open more space for empathy and care for those we oppose? Shouldn't this mean that more of us are kin?

Toward Conclusions

Why am I telling these stories? I have dozens, maybe hundreds, more tiny stories. We all do. There are not enough pages to tell all the stories of climate changing in our writing program alone. Yet perhaps I have already told too many—it is just so many layers of minutiae, isn't it? Deep in the sticks of details that merely index a vastly more complicated climate that is impossible to fully grasp or convey. I suppose that is the point. Institution transformation is often an exercise in recontouring the mundane, attuning to and resisting the every-which-way-you-turn nature of racist and other toxic structures in institutional life. Certainly, it is vital to sometimes embrace the uplifting idea that a million mundane collective actions might actually accumulate into a palpable climate change (though you may not be the one to benefit from or see the consequences). High impact, large-scale structural changes always matter, but my experience with institutional transformation work is that those kinds of changes are always the unpredictable end-result of ongoing atmospheric work accomplished by many in the tiny crevasses and rhythms of everyday life. This never-ending labor of equity work: it always happens on poisoned grounds, creates its own forms of harm, and its fruits are too often unknown or seem to evaporate in a poof, without trace. And it's exhausting. Knowing this yet persisting is an ethical orientation and praxis. Not one born of idealistic hope, but some sweaty, messy, I guess we must say human, commitment to the possibility for collective change toward X (justice, inclusion, a sustainable planet, etc.) while knowing there is no agreement on what this means, no endpoint, no eradication of violence, no process or procedure available to finally arrive, no arrival. Knowing this, and yet persisting, is not nothing. In solidarity.

Microclimate 5 (John and Candice)

We wondered: what might it mean to put our work in conversation in this space of collaborative thinking and feeling?

Hard to say, exactly. It's a coming together of forces. There's friction, for sure, energies pulling together for a moment before popping apart. Our chapters occupy different places. Bring them together, and you find yourself right here in a place where violence hangs in the air. We breathe it in; we are responsible for it, to it; it harms us. In this place, there's also some inchoate hope, you might say, a desire for respite and community, an orientation toward futures that could be otherwise, for remembering and creating new stories and ways, for pragmatism and praxis, for the doing of alternatives. It is dark and poisoned here, yet there's soil for dreaming about alternate collective existences, for asking questions about justice and a better place.

We speak through this, around this, to this, in what comes . . .

Candice is weathered. You're living in the present tense of institutional weathering that seeps into all you know, the ever nipping-at-the-heels that leads to a winnowing of the body and the spirit, first the blank look, then the cold room, *oh, upon review, we've reduced your budget*, followed by, *we're not funding justice work right now*, which is just one of many pastel colors of racism fading into the white noise of public research universities not caring enough to alter the apparatus, much less meet you for coffee, ask about the kids, ask to learn what you know better than they do about surviving a residential university and city. Hell yes—I know what you mean when you speak of the "active kicking back into place happening every which way you turn." I've been kicked some. I am reminded of Walter Mignolo's wry comment, "If you apply to get grants or fellowships to engage in decolonial praxis, be sure that you will not get them."[1] But we can keep poking with a broomstick, as you say, the "bridging from below."[2]

I'm not so sure, John, am I weathered? Or weathering?

> *I am weathering:* I remain here, pushing on, doing stuff, chipping away, tinkering, putting my back into it. I channel strength and joy from community or sheer endurance or from places I know not. Keepin' on with my keepin' on. I withstand it.
>
> *I am weathering:* I'm tired, depleted. My health has suffered. Sometimes, the waves flowing through knock me over. There seems to be less bodily energetics available to conjure or give away. Perhaps, I'm fading away, wearing down, soon to be vapor. There's less of me in me, somedays; I cannot stand it.

We withstand. We are worn.

We are all weathering unevenly in this climate. You show us this, John. Violence permeates the climate, you say, harming some more than others, bringing some in closer proximity to danger. There are cumulative impacts—could be overt or tacit harm, environmental stressors, or toxic remainders that enter bodies; alter them; change brain chemistry; edging some closer toward sickness, death, anger, violence.

Black weathering is now a matter of public health, once it was recognized and case-built into a traceable issue. For all the hideous direct violence by way of fire and gun, Shamard Charles writes also of the "slow crumbling of our health due to systemic inequalities and generational trauma." Black weathering is not ours to know. Your weathering is not mine, but if it descends to "DNA methylation and telomere shortening (premature aging of cells)," then the odd, sad common we share is weathering, generalized as such, but never universalized for all, as Liboiron would say.[3]

"It turns out the core mechanisms of writing programs and universities are coalescing everywhere to perpetuate a climate of racism, ableism, colonialism, and inequity." Yes, that's right, Candice. The core disciplinary structures undergirding writing, communication, literacy—they all have long helped to structure racism as a core medium of the university. But believe me, the climate-changing, antiracist work you list and the stories you tell of solidarity in the face of disbelief have brought decoloniality and antiracism into a different fold of the university.

John, mostly I see only the ocean of what remains to be done in comparison to whatever comes from my little broomstick jabs (and what a privileged stick, at that, held by a tenured professor with institutional power).

Fine, Candice, but we might remember that workplaces are surprisingly inimical to the very work needed to make them work better. Most workplaces, universities included, haven't yet processed that being antiracist isn't the same as being "not racist." (Beware of anyone who says, "I'm not racist.") Some knots get tighter the more you pull the string to untie them. Remaking the place you occupy means recognizing your own complicity in a problem you may not realize you're helping to exacerbate. Sometimes even well-meaning attempts to "fix" problems or to find "the solution" are themselves what make it worse. Yes, we see that; thanks for saying that so clearly. There is no end to the good work. And that is weathering. Weathering has nothing to do with giving up and folding under. It has to do with knowing the life wrapped up tightly inside institutional malaise, and the long durée of casual racism, classism, sexism for profit. To preserve Black life, we are told this: "Protect your peace at all costs. Don't sweat the small stuff. . . . Life is too short to let society weather our health and spirit away."[4] We can do this.

A pulling together: We can weather. We must. Try to create climates that do not weather so many, so unevenly.

So much pulling together and popping apart. Together, we wonder: What is the residence time of both love and violence? How can we measure the places in the body that are permanently healed by kindness or that are forever scarred by hatred? How might we perceive and locate the sanctuaries of life where particulates of care and connection hang in the air, as nourishment to breathe in, the places in the earth, the holds, that contain the ephemera and forgotten remainders of all the beauty ever created, evidence of ones who were held so dearly in life and after? And how do these sanctuaries comingle with indelible violence and evidence of horrors long past that persist within our institutions or our everyday lives or as molecules dissolved into the constitutive particles of life—sodium, oxygen, hydrogen, carbon, calcium, phosphorous, and iron—particles released from bodies vaporized upon impact or incinerated or liquefied or decomposed, and then returned to earth or ocean or breathed in as toxins, fleshy mist, molecule, and ash.

Perhaps, this final turn feels jarring and out of place, but it's the signal that you are about to enter a biological commons, an ossuary, an archive breathed in or buried, evidence of ecological violence—origins untraceable, forms innumerable, impacts immeasurable. We're all connected within these commons of violence and violent remainders because we created this violence, because we are all responsible for and touched by it. And, because we all, each of us, all that ever were and ever will be, return to and emerge from these commons, we are of them. You ask us to look, John, do not turn away; you say, really try to understand, breathe in this commonality. Go ahead and ask your questions of change, action, justice, a better otherwise, but only after you see this and breathe it in.

The climate to come is hard to be in, hard to breathe in.

Still, we are here; we're breathing, bearing witness to these commons, grieving together, in kinship, grateful to be, weathering.

6 Violent Clouds, Ashen Memories

(John)

If you work at a college or university that receives Title IX funding, then those institutions routinely announce their commitments to equal access and equal protection. Coterminous with those voicings, you may notice campus and community announcements of crimes committed or potential harmers on the loose. One kind of announcement is timed by order of legal statute, the other by order of incidental occurrences worthy of public attention. Incidence reports of crime, mayhem, and the threat of either now coincide with campus and city policing to be nearly synchronous with the events themselves, thanks to digital media and bystander reporting. Fear and harm lean ever the more into each other, as both campus and city offices diligently upgrade violence prevention programs and assessments to deepen a shared infrastructure for maintaining safe working and learning environments at least for those most likely to be protected from harm. Whether by human action or natural expression, ecosocial violence constitutes one weave of the present, reported incidentally as biological disturbance (fires, floods, unruly parties, petty crimes, and domestic violence are all biological in their way).

This early warning system was severely tested on December 30, 2021, when the Marshall Fire, driven by one hundred mile-per-hour winds, swept down the Flatiron mountain uplift and across the high plains around Boulder, Colorado, to incinerate one thousand homes leaving a debris field of scorched earth, twisted metal, smoke, stench, and dust that settled into residential neighborhoods and businesses near where I lived. And then came an early, welcome blanket of snow on top of scorched earth and glowing metal. And then came the accusations of system malfeasance (even though I live within one of the wealthiest, most climate-attuned, and data-informed communities in the United States). And then came the mourning: first shock, then moans, then sleepless nights in forced migration, with pensive looks at the sky and weather reports. I begin this essay by pointing to the aftershocks in an assaulted community. Despite the best of intentions, data mills, and environmental sensors, gaps appear, dangerous ones, between what institutions are required to do, what they aspire to do, and what they fail to do to mitigate biological violence when it compounds atmospherically.

In terms of sexual violence, my campus, like other federally funded institutions, must legally and financially uphold three overlapping policy registers: Title IX from the 1970s outlawed discrimination on the basis of sex; the Clery Act required full disclosure of actual crimes and potential risk some twenty years later; and more recently the Violence Against Women Act, signed into law by President Obama, amended the Clery Act to sharpen prevention and transparency in reporting domestic violence, dating violence, stalking and sexual assault.[1] A mundane Clery incident report, barely more than a paragraph and a blurry image of a suspect, prompted this essay in fall 2020, because the report identified a curly headed, light-skinned male as Black and menacing because he brandished a knife while being confronted by a group of young (White) men, who were loitering in a campus parking lot during full COVID lockdown. What struck me about this flicker of an image was the flimsy connection of danger to race while ignoring the potential harm of loitering Whites, but then also that this Clery report had nothing to do with sexual assault or domestic violence and with no mention of extenuating circumstances or violent conditions on my campus.[2]

Obviously, such warnings cannot account for the scale and breadth of what so many women have told me over the years that a climate of violence exists in and around campus (due to loitering, mostly White men). My university, like so many others, has its own Brock Turner story,[3] the Stanford swimmer convicted of felony sexual assault only to be sentenced to six months in jail and probation. The crime, the incident reported, the enactment of justice, and even the media accounts of the harm Turner brought to the victim and to his community, fail to capture the ephemeral nature of violence that precedes and exceeds the criminal act or its adjudication. The Clery Act exists today largely because the parents of Jeanne Clery, a first-year student who was raped and murdered in a residence hall at Lehigh University in spring of 1986, pressed for legislation that requires campuses to gather and report crime data, with the belief that increased awareness by yearly statistical and incident reporting will prevent crime.

Yet as I studied the Clery Act as policy, I could not ignore the billowing, compendium of biological violence pinned to heightened tension around racist policing in 2020, twinning all the more with fires, tornadoes, and floods, and governmental malfeasance. My pandemic nightmares were rooted in material disturbance such that "atmosphere" identified much more than a "rhetorical ecology" or the structured feelings that seep into one's soul or amassing in the street and in the climate on a given day.[4] The Clery Act indexes an actual harmer, Josoph Henry, and his friends who attended several fraternity parties on the "Hill" at Lehigh, drinking and becoming increasing agitated and aggressive. Henry's friends, later turned informants in the courtroom, reported their aggression to be normal "because they (as men) were all angry and 'acting the same way.'" These friends also conversed with Henry about "the difficulties of being black students at Lehigh." Henry was punished for his crimes, but at least two conditions preceded the incident, looming as atmospheric violence. Henry's friends encouraged aggressive behavior through thinly veiled endorsements of male dominance. Jeanne Clery apparently said "no" to his advances, but he was told by his friends that "he shouldn't let a woman get him so upset" invoking a tolerance for assault cloaked as friends who partied on the Hill.[5] It appears then, as it appears today, that White male aggression fuses with socially palatable racism.[CR] In this case, such racism is underwritten by the story of a Black man

---

CR This question resonates. It's the kind of question I find myself asking, too (and it's one I've often felt ashamed of or shamed for asking). I hear you urging us to acknowledge violence as inescapable, indelible contour in the climate, tangled up in our ways, our relations, in the all. I hear you urging us here to ask about and address the violence done to women always. And also to ask: What happened to this young man? What forms of violence might he have experienced that contributed to him arriving in this moment of causing such harm? I'm reminded of Nadya Pittendrigh's important rhetorical scholarship on restorative justice. She asks about these things, too, and about how our punishment mentality might be in tension with our justice aims. This line of questioning asks us to wonder how we might be eschewing our own complicity or missing the possibilities for keeping someone in tow as "one of us," even as we hold accountable or punish and never tolerate. She asks, like you: How might we all be involved in this crime? How is the community wrapped up in this pain or in the conditioning of this place of violence? How might the community help heal these wounds? How might we keep more of us in the fold of our humanity and our relations of care, even those who do harm? It isn't popular to ask these things, is it? I know, it is not

with no prior record, a former honors student, who struggled with alcohol and was caught up in dangerously White, patriarchal conditions.[6]

Toxic masculinity appears in these cases, as it does in all cases of racist violence, to exceed the temporal, physical, and geographic boundaries of an incident,[7] to gather its energies then dissipate its forces in something like a regional weather event, alongside and within deep sinews of ecological violence that apparently were part of the life and death-sentence of Josoph Henry. Toxic masculinity, toxic racism, toxic indifference, and dangerous proximities coagulate as a dream, a condition, an atmosphere beyond the empirical record, a confluence of viral agents comingling biologically and sociologically to upset simple-minded causality. To admit that violence is "in the air" has to mean more than dark clouds in the sky or soul, more still than "cruel optimism" or melancholy and certainly more than stacks of data that underwrite someone or thing to blame.[8] *Weathering* is now an accepted term in behavioral science for the environmental hazard of living a Black life, as told by Shamard Charles, a death by "repeated exposure to socioeconomic adversity, political marginalization, racism and perpetual discrimination." All of Charles's figurations are ecological, not merely sociological, as they speak to atmospheres, particulates, and weather:

> Racism forms cracks in our spirit, like cracks in the pavement of a busy road. Constant bouts of discrimination fill and expand the cracks like raindrops. Over time, the crack becomes a pothole that no longer resembles its original form. The same is true of our cells over time. . . . Stress, self-doubt, anxiety, and fear causes DNA methylation and telomere shortening, signs of aging in our cells . . . a biological thumbprint of generational trauma on our youth. . . . When Black people say they are mentally and emotionally tired, or that they feel the weight of the world, believe them.[9]

There are those, of course, who would refuse the historical, biological, experiential lessons of Black weathering,[10] foremost because racism serves their interests, but last year, and right now, and for the near future, I turn to such declarations as they speak more clearly to a pressure I live with each day, although clearly my history is not Black history, my city is not Minneapolis. "Cracks in our spirt"

---

always the right time or circumstance to ask. Asking these questions can feel like a betrayal or violence. Still . . .

indexes for me the last year of Trump's presidency that fronted a campaign of fear and lies, stoking a national tolerance for police and border brutality; science denialism around the climate and the pandemic; the wrecking of democratic institutions through voter suppression and election subversion; and a political, electoral investment in BIPOC suffering hallmarked by the deaths of Ahmaud Arbery, Breonna Taylor, and George Floyd in a matter of days, with so many others, before and after.

The year 2020 will forever be known as a prime number among all years of Black death as it also dates a historic fire season across Colorado, with seasons of smoke and polluted air from fires across the western United States and Canada that lasted well into 2021. "Cracks in our spirit, like cracks in pavement," and for me and my neighbors, cracks in scorched earth and cracks in our faith in institutional resilience. The year 2021 began with an assault on the Capitol by White supremacists followed three months later on our "Hill" by a freedom-starved party (their depiction) of eight hundred White, privileged students that morphed into a destructive riot.[11] Two weeks prior, we had a mass shooting in a grocery store where I often shopped.[12] The "carnage," as a *New York Times* article called the groundswell of gun violence in the United States, might then be rightly called an *atmosphere*, and take its toll as *weathering*, leaving as residue a mist, a film, a poison, an atmosphere with little to no separation in time and in place between one biological event and another. Smoke and toxins from Portland, Oregon, protests mingled with smoke, dust, broken glass, and blood mist in Boulder. Sagging beneath the phenomenon, the article tried to capture, gun violence (like racism, like sexual assault, like climactic catastrophe, like political insurrection) is difficult to see because "the shootings never stopped. They just weren't public" and so they too settled into the cracks in our spirit, city, and soul.[13]

The maps, the lists, broken bodies, broken glass, the burns, the floods, the contaminated air, soil, and water that intensely accumulates as an unspeakable force near my doorstep and then yours, and then in the next *terroir* and metropole, speak to a world of ephemeral and enduring histories with far-flung particulate remainders—filling in the cracks, making them wider and deeper, down into the cellular structure of trauma. I have searched high and low for some sort of public record, some meteorological report that overlays fire, a *demos* on the run, death, and protest to calm my nerves by recuperative art, reducing in some sketchy, loving way, the imperceivable.[14] I desire some kind of expressive measure of the elemental constitution, flow, and dimensionality of pluri-violence (let's call it that!) and the ensuing in-permanence of its wake, as I have come to know that strict causality is an illusion.[15]

So, I found myself staring at interactive maps of satellite data to find an accurate representation of the gregarious, unrelenting animism of ecological duress, with sinews of raced, sexed, and gendered death folding into the water and air in the lost lives, habitat, and sanity of my neighbors. The texture of that map would be deeper than precarity, more than the residues of extractivism, stickier than melancholy, a texture closer in kind to all passing things. The following "rapid refresh" atmospheric map from the National Oceanic and Atmospheric Administration (NOAA) captures the "emissions and transport" of "near surface" particulate matter afloat on July 22, 2021, centering the United States but with Earth in near relief. The color, scale, and drift are truer to my fears because the world's atmospheres are on fire and merging, and beneath the purple plumes in the intermountain west are the violence and protest of that summer, otherwise known as weather.[16] Staring at such images assuages my fear with fantasies of atmospherics, embracing a weathering that endures in some rooted, particulate way as commemoration but never ceases to infect those people and machines that exert sovereign control.

Speaking only for myself while embracing my community and others, I cannot wait for police reform, ecological responsibility, economic justice, responsible leadership, or racial tolerance, any more than Shamard Charles waits for Black weathering to clear. It will not clear, and so the only reprieve close to the breath and taste of life would be found living within violence and not outside of it because "bare life" was never, will never be, meted out equally and irrespective of race.[17] Any such bareness would reside in the afterlife of violence. I cannot read Intergovernmental Panel on Climate Change (IPCC) reports, or the catalogues of mass murders, or the rising tide of White supremacy talk, without some medium and measure for the constancy of death, destruction, and decay that promises nothing more that proximity, residual habitation, and a dream.[NS] I held these thoughts right before the strange winter winds whipped

---

NS I feel the aching in your words. It helps me think, reflecting across the chapters, Candice's in particular, that something needed, something I would benefit from at least, is a practice of living with the constancy of pain, of collective loss and trauma. More than enduring, or weathering, a practice for standing in the inferno as we'll get to in the wrap-up conversation, of more than *withstanding* the intensity and globality of pain and death. Death is

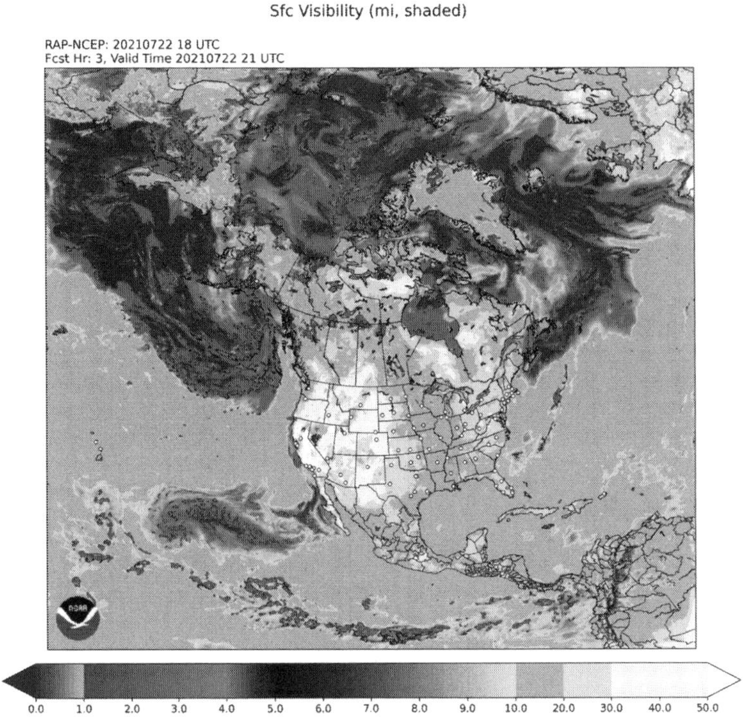

Sfc Visibility (mi, shaded)

RAP-NCEP: 20210722 18 UTC
Fcst Hr: 3, Valid Time 20210722 21 UTC

0.0 1.0 2.0 3.0 4.0 5.0 6.0 7.0 8.0 9.0 10.0 20.0 30.0 40.0 50.0

FIGURE 4. RAP-Smoke Model Fields, daily map, July 22, 2021, from the Global Systems Laboratory, National Oceanic and Atmospheric Administration, https://rapidrefresh.noaa.gov/RAPsmoke/.

along the Colorado front range on December 30, 2021, so quickly that every dimension of time, anticipation, and security perished, backlit by red skies of pending doom at midnight.[18] I revisited these thoughts in the weeks to come while a mid-twentieth-century-style ground assault rolled into Luhansk Oblast and all because, once more, a paranoid cis-White man pined for lost days of ethnic mastery.[19]

a hyperobject in Timothy Morton's terms. I am not a spiritual person; I know many turn in that direction, but I think a practice grounded in exposedness, not divinity or visions of life as antideath, is required and is what many are trying to imagine even if they do not see themselves as saying exactly that. And I think it has to do with rhetoricity, but not the rhetorical tradition as we all were taught.

An Alterlife in Residence Time

Michelle Murphy coined the term *alterlife*, to release from captivity the inner vitality of a "life already altered, which is open to alteration" as a chemical relation of belonging by way of a molecular community: "Alterlife embraces impure and damaged forms of life, pessimistically acknowledging ongoing violence" and in doing so refracts more clearly the spectral forces of the "technoscience" that infuses everyday life with pollutants and degradation on a global scale.[20] To witness this alterlife, one has to enter the scene of molecular dissolution and an ongoing distribution of particles in media like air, itself once a cloud, once an ocean that moves across a city, thickened by smoke and dioxides that clings like dew, like blood, to the skins of buildings and the pores of people only then to drip down to evaporate into a silent drain of tributaries made of gutters, sewers, and creeks. Body and mass come to rest in the singularity of alterlife so molecular, so fine-grained and thin that spiritual or communal relief are found in confinements to a molecular solidarity: "one cannot simply get out" of an alterlife although there are legions upon legions of disciplining forces that would bet their epistemological existence on their instruments that stop our descent into alterlife entanglements.

For someone like me to enter the alterlife, I would first have to admit that I was already there, and that those I scorn or cherish were already there as "guest and settler, caught up in the chemicals and the water." I would have to imagine a particulate solidarity beyond the human "self" and all of those received notions of community or state, and if that jump were made, then my world might "become something else, to defend and persist, to recompose relations to water and land to become alter-wise in the aftermath."[21] This alterlife is resolutely "autobiographical" if the reflexive subject can tolerate an autobiological continuum, below and within the technoscientific manufacture of crime, death, loss, and decay. For Murphy, the life lived ordinary leaves a stench, a mist that sluices down built and natural surfaces, into the sewers, out into Lake Ontario where there is no escape from polychlorinated biphenyls (PCBs). As chemical we dissolve into the rust of technoscience. There can be no shelter, no antidote, for chemical contamination, nor for viral hatred, death, and loss, not without some embrace of this liminal, molecular stew and the solace to be found by belonging to the remainder.

What I propose in this chapter is both radically empirical as it is anagrammatically expressed. There are instruments, metrics, thresholds, and archives to consider that provide degrees of rigorous depth and measure to ecological violence—an entire techno/social scientific technicity of clocks, counts, parts, and accumulation. Clearly, the past years where I live has been itemized, demarcated, and weighed mostly as the ecological afterlife of bark-boring insects, deforestation, fire, and floods that could be known as "elemental media" so vitally resonant with Indigenous thought and so discredited in the analytical canons of disciplinarity.[22] Empirical records become radical, I propose, when biological violence against bodies and matter spreads out into the widest and wildest diagram to resolutely demonstrate that violence will always be concomitant and unrelentingly atmospheric.[23] Such radicality leaves divergent elements to circulate in a dis/passionate, biological repose, the aftermath not merely as threatening weather but as residue as blood, tissue, fragments, and dust that settle into a molecular containment and release. Neither formal incident report, nor statistical tally, neither number nor figuration, contain the scale and substance of this kind of weathering accessed foremost by breathing and taste. And, so my writing, at times, aspires to what Christina Sharpe calls the anagram,[24] an abrogation of form and syntax, arrangement, and disposition, to allow "new meanings [to] proliferate. . . . As the meaning of words fall apart, we encounter again and again the difficulty in sticking the signification." And so, my writing tries to invent a phantasmagram appropriate to the violent encounter and its aftermath closest to home,[25] but also that grants the presence of "imaginaries and hopes [that] were already there, circulating in excess" well outside of calculated systems of loss and death and aided by imperceivable disruption.

To write that way, for me within my raced, gendered, sexed, aged, and regional limits requires that I set aside many of the formalist practices of colonial rhetoric that specify what constitutes the body, the thing, the place, and the word,[26] that tend to disavow the radical empiricism of truly weather contamination.[BM] We don't remember gunshot victims for their molecular mist.[27] We,

---

BM Michael Lechuga's article you reference here has become a touchstone for me. I love this article. Lechuga points to the radical transformations of knowledge, and especially for theory, method, and pedagogy, that would constitute an anticolonial approach to rhetoric. For him, such an approach is community-based and "privileges the voices and actions of

as a society on the run, tend not to remember broken cities for their molecular dust. We sometimes remember fire and annihilation for its smoke and ash as a biblical ritual and ethnic annellation, but those legacies tend to ignore the fire, smoke, and ash born of technoscientific, climactic destruction.[28] The print-wrapped incident is perhaps one of rhetoric's oldest yet unexamined modes of colonial rule as they deter passionate, fleshly recall. Decolonized, each instance of systemic violence could reach beyond an econometrical rendering written and spoken anagrammatically as phantasmagoria. When something breaks—a neighborhood, trust, economic security, domesticity, pastoral tranquility, and so forth—the rhetorical move seen over and over in public life entraps the event in a written report as the case, the byline, and the soon to be archived and forgotten.[JL] The incident forecloses the inevitable dissolution of life into its actual homeland, its alterlife, and so forgets a different scale and mediation of remembering by way of a different ossuary as the most common restive place, the final remainder and an origin of becoming.[29]

The rawest edge to catastrophic violence is that it is metered unevenly by way of racial and economic territories and embodiment, and the incident report is one of many "settler technologies."[30] Taken as an arsenal that includes legal and policy documents, economic profiling, social embattlement, and killing machines, they constitute what Jason De León calls "necroviolence" with an insidious "outsourcing" of violent means and causation to "animals, nature, and (ordinary) technology."[31] Don't fixate on "natural" processes of decay as in "ashes to ashes, dust to dust"; rather, look for animals, nature, and technology infused to install a masterly infrastructure as both an instrument of violence

---

those in vulnerable communities meeting power head-on and asks scholars to deprivilege their assumptions about academic knowledge. It is collaborative and not built on subjective differentiation between researcher and objects of study" ("An Anticolonial Future: Reassembling the Way We Do Rhetoric," *Communication and Critical/Cultural Studies* 17, no. 4 [2020]: 384). I think this is so essential and yet also so hard because of our entrainments, "naturalized habits" in hierarchies of knowledge and the infrastructures, climates, that perpetuate these ordered distinctions within academia. But he also models how to attune to and disrupt these habits and structures, through reflexive consideration of complicities and histories for how his own scholarship has changed through time.

JL Part of this forgetting is denying how our bodies are all necessarily archives, possibly because the process of archiving these catastrophes in ourselves produces forms of disability. The call to reject ableism and instead look for forms of the social that welcome multiple forms of ability might also recenter these systemic violences, refusing the violence of negation.

and a chrysalis room on the way to the ossuary. When Saidiya Hartman speaks of the "afterlife" of slavery, she enters a troubled room unknowable to me (perhaps more to you), and certainly to colonizers and enslavers, anyone removed from the cutting edges of necroviolence against her people, her race, and her economic caste.[32] There are only "scraps" of those years and places that gather as historical and literary remembrances in the manifests, the ledgers, the inventories, bills of sale, and itemized lists, the logs and diaries that piece together the long *durée* of calculated death and decay of Black life. The scraps paper a room, as a hold, an ossuary of uneven proportion and entry because the "past is neither inert nor given."[33] Christina Sharpe, indebted to Hartman and Dionne Brand, also knows this ossuary: "The rooms are not empty and the scraps are what we have to offer."[34]

Scraps as tattered reminders, as fragmented records, shattered lives, and body particulates—claiming them requires a willful entrance into an alterlife, and for Sharpe to embrace "those Africans who were in the holds, who left something of their prior selves in those rooms as a trace to be discovered. . . . they, like us, are alive in hydrogen, in oxygen; in carbon, in phosphorous, and iron; in sodium and chlorine . . . they are with us still, in the time of the wake, known as residence time."[35] Residence time as a volumetric in the ecological sciences measures the rate of exchange in a volume (as small as a glass or as deep as an ocean or as vast as a desert) for any substance to arrive at a steady state of dissolution. A trace, a fragment, a body, a drop of blood reaches a point of equilibrium through a process of "organisms eating organisms" within a watery ossuary where Black bodies finally escape harm "cycling like atoms."[36] Residence time calculates the duration of entrance into a biological ossuary wherein the sea is first surveyed as an economic zone, then honed as an instrument of necroviolence, then twisted into an alibi at the point when Black bodies break the surface to drift down into the wake of Black history because, as chemistry allows, "human blood is salty, and sodium . . . has a residence time of 260 million years."[37]

And what of arid killing fields strewn with bodies that stumble, resting for a moment, then stumbling again to catch a speck of shade only to faint and then slumping into the heat to rot. Jason De León's oceanic ossuary is the Sonoran Desert, just south of the U.S. border which like the enslaver's sea has everything to do with a merciless, deliberate calculation of time, weight, food, distance, heat, water, the thinnest of shelter, and a waiting to die with scavengers (as "organisms eating organisms"). For De León, necroviolence exceeds the momentary effects of asphyxiation and dehydration by extending the "generative capacity for violence."[38] Maritime enslavers for centuries profited from a model that increases economic gain by minimizing the cost of transit. The policy of the U.S. Board Patrol doctrine called "Prevention Through Deterrence" employs the desert to kill by exposure 45 percent of the "2,238 dead migrants examined by the Pima County Office of the Medical Examiner between 1990 and 2012" and with another 36 percent "too fragmented or decomposed" to determine a cause of death.[39]

Residence time for a migrant body in the Sonoran Desert, falling parched into a coma then death, has been calculated carefully by tracking the decomposition of pigs dressed like migrants who paused, as they do, beneath a sliver of shade. The tissue and organs of a pig rots similarly in substance and time of decay as to those of a human that translates into a "taphonomy" or the social processes of death as a "combination of human and nonhuman elements that impact biological remains."[40] Migrant census data captures none of this, but residence time may point to a time and place of molecular singularity of this medium's diagrammatic and temporal spread. In the desert, the final resting place is typically under a tree, where the fallen body is "rotisserie-cooked by the rotating sun."[41] Skins discolor, organs balloon, fluids drip, maggots chew, and the body contorts as packs of crawling and flying scavengers patiently wait to commence their in-step, in-kind "crescendo" of ripping and tearing and scattering the body so that within a few weeks "a person left to rot on the ground can disappear completely."[42]

I have come a long way in this essay from scorched earth to bloodied produce, from campus violence to civic insurrection, and now to oceanic and arid ossuaries. As I see the "wake work" before me, and for my community, there is respite in knowing that bodily violence descends into a particulate remainder. The memory theorist Avishai Margalit proposes that "ethical" remembrances depend foremost on "thick" relations with those of the "parent, friend, lover, and fellow-countryman [*sic*]," while "moral" remembrances belong to the "thin" relations with the "stranger and the remote."[43] For his Jewish kin and in particular for those who survived apartheid European life, ethnic cleansing, and forced migration, Jewish thickness would have a tensile strength. Yet, thinnest of relations belong to bodies at rest. Sharpe proposes, "We must think through containment, regulation, punishment, capture, and captivity and the ways the manifold representations of blackness (at times folding into brown, red, migrant, servant, poor, woman, queer through compassionate dilution) become the symbol, par excellence, for the less-than-human being condemned to death."[44] The proposal here is to reimagine ethical, moral, and spiritual work through radically empirical pathways to an alterlife of thin relations that require "enfleshed work."[45] I can only read about the afterlife of slavery and then turn to count its remainders in my neighborhood. What I can also do is learn how to make sense of the choking fear of a fiery death and the rancid smell of gun violence. Sharpe does not intend this, but she comforts me by saying, no, not yet, John, you cannot go where I've lived, because the "hold cannot and does not hold even" in the ship, in the desert, in the grocery, in the neighborhood. Her thickness differs from mine, but my thinness cries for "antiblackness as total climate" that might afford a safer day.[46]

Thin Relations of Dust and Air

A technoscience that brazenly employs biological dissolution to harm and then hide unwanted bodies, a necroviolence that assembles the mechanics of ships, walls, navigation, and surveillance with land, weather, water, and air—all of that apparatus-building and shape-shifting fuses racism to infrastructure. Yet beyond the fortress, an atmospheric remainder moves, thin as air and mist, to recapture the taphonomy required to correct the ledger and to sponsor phantasmagoria of national sovereignty and economic futures. That appears to be true for Katherine McKittrick who employs taphonomy, as the "study of decay," and "diagenesis (the changes that take place after the final burial)" to resurrect Black death as an externality of plantation life thereby insuring that the city's racist infrastructure retains its "migratory" tenacity.[47] Her "plantation futures" track "the plantation toward the prison and the impoverished and destroyed city" to honor those who died "under bondage" to keep the plantation alive as "an ongoing locus of anti-black violence."

Phantasmagrams rewrite history not by canceling the razor edges and numerical precisions of technoscience. Rather, as Michelle Murphy explains, they echo technology and science dreaming, as they would, about "post-colonial futures" as "new worlds of rationalized equality, worlds without racism, words of industrial prosperity."[48] Technoscience has its own dreams of a global caste system with "precarious life as a kind of surplus, a devalued and unwanted excess amenable to erasure and optimization." The phantasmagram, thus, "does not undue objectivity" so much as show that numbers dream in their ways as "wishful calculations" for colonial futures built upon the "palpable sense of the large immaterial forces that (data) models aspire to glimpse."[49] Phantasmagrams bedevil the numbers to delink the colonial dream from dreams of escape by tapping into the "many wells of unaligned dreaming that capital fails to register,"[50] another kind of incidence reporting all together.

There is nothing more thinly moral that breathing someone else's air, which now can be imagined in this essay to float (as it actually does) from the fires and protests in Portland in 2020 to the pollution, fires, and protests across the seasons in 2020 and 2021 where I live. Tear gas, blood, broken glass, and hatred spiral out across over five hundred sites in the United States, and well beyond that globally, to make Black Lives Matter probably the largest social movement in U.S. history.[51] Hatred and violence, pollution and contagions not only change the weather, it appears the seasons depend now upon them, and as established earlier in this essay, it does so unevenly.[52] The given story of an assault on our sovereign soil in the United States describes terrorist motives and means resulting casualties of 9/11 because radicalized Mideastern Muslim people resented U.S. imperialism, so they hijacked commercial jets to fly them into monuments of wealth, military might, and governmental control. The phantasy does not reject the facts of what happened so much as displace the locus of remorse and lessons toward future collectivity. In "Poem Written after September 11, 2001," Juliana Spahr looks way past unscaled catastrophe and political posturing to find a common molecular stratum and a common moral complicity.

> There are these things:
> cells, the movement of cells and the division of cells
> and then the general beating of circulation
> and hands, and body, and feet
> and skin that surrounds hands, body, feet
> This is a shape,
> a shape of blood beating, cells dividing
> . . .
> In this everything turning and small being breathed in and out

> By everyone with lungs during all the moments
> Then all of it entering in and out
> . . .
> How connected we are with everyone.
>
> The space of everyone that has just been inside of everyone mixing inside of everyone with nitrogen and oxygen and water vapor and argon and carbon dioxide and suspended dust spores and bacteria mixing inside of everyone with sulfur and sulfuric acid and titanium and nickel and minute silicon particles from pulverized glass and concrete.
>
> How lovely and how doomed this connection of everyone with lungs.[53]

There is much more to be said about Spahr's verses for how they invoke the spaces within and among hands, rooms, buildings, neighborhoods, cities, regions, nations, continents, islands, oceans, tropospheres, stratospheres, and mesospheres, to scale outward, then upward, then downward, then inward, but in this section, I wish to spotlight breathing in and out after moments of racial and ecological violence, as does the poem, and with dusty, smoky contagions.[54]

Glenn Albrecht and his colleagues in environmental and ecological health coined the term "solastalgia" to identify "somaterratic illnesses" wherever and whenever homelands are broken down then contaminated such that biological and artificial extremities become nonrecognizable and the air more toxic.[55] While nostalgia sets in when a person longs for home, solastalgia sets in when someone's sense of place is shattered—quite literally those sensations given to sight, smell, space and breathing. Albrecht's research team studied mineral extractivism and damaged atmospheres thereafter—the constant pressure of vulnerability given to persistent drought in rural Australia, and "open-cut mining and power industries" that led to respiratory suffering, in particular for women.[56] The scars, dust, and debris from open-cut mining fuse with other waning natural resources (water, fossil fuels) and then roads, noise, and contaminations from older ways of life. Extensive drought heats the air while weakening the social and natural defenses in a region—taken together mechanized violence treats everything as opportunity and as a victim. As one informant reported to Albrecht, "We're coming into our fifth year of the drought. . . . our gardens have had to die . . . so it's very depressing for a woman because a garden is an oasis. . . . that's all gone. . . . you've got dust at your back door."[57]

Across the twentieth century, there is no more notorious machine of necroviolence than the ovens of Auschwitz, Buchenwald, and Flossenbürg that epitomize systemic ethnic and racial hatred by employing the elemental media of fire and its particulate remainder as dust at your back door. Topf and Sons invented "incineration" not "cremation" chambers to allow the company to market the ovens for clients all over Europe. For Nazi SS clients, the ovens were further refined for the quickest burn cycle and the highest volume to render the ashes "unidentifiable and intermingling" to then be mixed with sawdust and general dust and stamped with false identification numbers for Jewish victims.[58] The smoke from the chimneys, the ash from the ovens were inhaled by the Sonderkommando, the Jewish slaves that worked the showers, the ovens, and the ash field. Ernst Israel Bornstein recalls that work:

> Here in Flossenbürg, I had to bring the corpses to the crematoria, and there I stood next to the ovens where I saw the fire and inhaled the sweet smoke of burning human flesh. We were filled with a mounting dread of death as we watched the burning corpses . . . they heaped the dead into a mountain in front of the building. . . . When there was no room left for more piles of corpses, a ditch was dug adjacent to the barbed wire at the edge of the camp. . . . We threw the corpses into this ditch, poured tar over them, and set them on fire. The rising smoke polluted the air throughout the area. Both crematoria burnt day and night.[59]

I pined earlier for dreamy, melancholic maps of drifting particulates over distant territory to show the spread of fire and grief in 2021 and to suggest, as the phantasm allows, a floating, gritty mode of commemoration. Perhaps as this essay unfolds, a particulate memory, held deep in the lungs and with a taste in the mouth, is all the more plausible.

Flight 93 was the last of the trio of planes driven into monuments to the economy, into monuments to military mastery, and then into pastoral life on September 11, 2001. Driving home from the grocery in 2021, twenty years after 9/11, I caught the last minutes of an NPR broadcast of *Sacred Ground: A 9/11 Story*. The show returns to the now memorialized site of the crash of Flight 93 into a Pennsylvania pasture and to comments from a resident who reports for a news outlet and who owns part of the land. He knows the local coroner as a friend to the NPR broadcast's writer and narrator. I had been reading accounts of how survivors struggled to breathe freely, twenty years later, because toxins enter the lungs and remain there, waiting in residence time.[60] I listened to soundbites on the anniversary: political speeches and news reports, and then on this reunion day, the voices of Debby Borza whose twenty-year-old daughter, Deora, switched to Flight 93 to hurry home and then Lori Quadragno who lost her brother in the crash. Debby was instrumental in staging two commemorations suppressed by powerful, national security forces—a fight to allow an unprecedented playing of the flight's black box and the struggle to have the crash site preserved as a burial ground for the families.

As a result of passenger resistance, the plane hit the ground at five hundred miles per hour and pulverized. The coroner drove to the site to find nothing, no fusillade, no noticeable debris, and after lengthy searches only 8 percent of the plane (a mere 650 pounds) was recovered as a jangle of ordinary clutter, cataloged later into a searchable list of hair bands, belts, pants, shirts, shoes—and then one fragment recognized by Quadragno, her brother's wallet. Nothing remained on the day of the crash but the smell of jet fuel and a crater wherein the plane disintegrated. "You could hear this melted plastic dripping out of the trees . . . (pissh, pissh) . . . You could hear it hitting the ground and sizzling."[61] Twenty years later at a commemorative gathering, Lori Quadragno shared the wallet and the smell of jet fuel that lingered. She skipped the black box hearing and avoided the company of other families, but on this day, she said:

> Ok, 20 years, so much time has come, and I'm at a place where the pain never goes away, obviously. I've never had closure in this story. I never will. . . . But when I touch these relics . . . I want to be there because also, I know how the wind feels (the crash site). . . . I know how that land smells. . . . I know what certain tree bark feels like to touch. I know what the soil feels like in my fingertips. . . . It suits my soul.

And for Debby Borza, whose daughter would have been forty today, and who moved from California to the East to guide the building of the memorial, Lambert the reporter and former landowner is the closest of friends with her: "She's lived and breathed this place for 20 years." Then walking the grounds, combed clean years earlier, he stops, looking at his feet, to find a fragment, a ten-inch, jagged, and ripped shard from the airliner.

> So, you have to imagine it was up in one of those tall trees. And just over time, the wind shook it loose. The rain shook it loose . . . or the weather—you know, again, the ground shifts and sifts. And it comes up. Amazing. Twenty years.

The scraps of the archive live on in the lungs of those nearby and those who rushed to help. Catastrophe on this scale and the ensuing fragments and dust do not wait to be assessed by an Air Quality Index. The pasture envelops the plane, as does the desert and the sea, for its decimated remainder, an ossuary that fuses soil, jet fuel, minor debris, and all other chemical additives washed down from twenty years of rain and snow. If there is purification, it comes from water and new growth releasing nutrients into the air. A few tattered relics form the archive, and as always, they are turned into a list. For the Holocaust, bodies are burned with new-found efficiency and thrown in ditches; toxins are forced into their lungs and those of the Sonderkommando to hide the list. Bodies scattered into soil because of ethnic and economic disparity can apparently turn a plane into plow. Bodies through a horrible, racist calculus are thrown overboard to drift down and dissolve into the ossuary of the sea. Bodies in the desert are so parched and eaten that remains blow away in the wind. Mining the earth scars the land with roads and debris and with air and water so foul that home in sight and in mind disappears into melancholy. "How lovely and how doomed this connection of everyone with lungs."

When I think back over the essays I've written and studies I've conducted, this connection of *everyone with lungs* appears everywhere I've traveled and worked.[JL] In Kent, Ohio, on May 4, Jeffrey Miller's blood puddled along the curb that would soon wash into a sewer.[62] The dust of ruination drifts across Cyprus, thirty years after war, to lodge deeply in the consciousness of the anthropologist Navaro-Yashin as "abjected" matter and as melancholy for the surviving informants who speak of feeling "suffocated in this territory."[63] And

---

JL So, is part of the next step to consider us all as bearing a form of atmospheric lung disease? What might it look like if we took a shared debility that exceeds the one body or human and centered that as the starting point? For rhetoric, we might think about delivery differently, not trying to sidestep the physical strain of breathing to speak but instead calling for attention to the beauty of the rasp, the clogged throat.

the air in Ciudad Juárez, Mexico, is thick with smell and taste of violence after NAFTA: "In this new way of life, no one is really in charge and we are all in play . . . and I feel this in my bones. . . . The violence has crossed class lines. The violence is everywhere. The violence is greater. And the violence has no apparent and simple source. It is like the dust in the air, part of life itself."[64]

In all these scenes of violence due to economic progress, ethnic hatred, the demolition of cities and farmlands, chattel slavery at sea, migration and exhaustion in the desert, a thin veil of dust in the air and watery fusion settles into an ecosystem. Land and sea remain defiant in providing an ossuary from which to consider a cold, dry truth about living with and beyond hatred and violence and with this reflection in "Ossuary I" by Dionne Brand.[65]

> I lived and loved, some might say,
> in momentous times,
> looking back, my dreams were full of prisons
>
> in our narcotic drifting slumbers,
> so many dreams of course were full of prisons,
> mine were without relief
>
> in our induced days and wingless days
> my every waking was incarcerated
> each square metre of air so toxic with violence
>
> the atmospheres were breathless there,
> the bronchial trees were ligatured
> with carbons

Aspiration Has a Line through It

It takes ten minutes for a body to drown in salt water, for the lungs to fill with a foreign object and therefore to aspirate. Drawing the deepest of breath often accompanies dreams of a better tomorrow, a better life. Educators cannot ignore the increasing numbers of students at school exhibiting more symptoms of anxiety, depression, and attention deficits putting their aspirations on hold because of workloads, bullying, or, what is often true in my classes, loneliness that can lead to a shortness of breath, heightened blood pressure, and the need to catch one's breath. They hyperventilate, as if they are drowning, which of course as a social metaphor, they are. Christina Sharpe turns to aspiration because, like Blackness itself, words possess an anagrammatical capacity to cross lines. We aspire to live; we aspirate and die, and for her in parallel, "blackness anew, blackness as a/temporal, in and out of place and time, putting pressure on meaning and that against which meaning is made." Aspiration "doubles" even "trebles" in Blackness as part of wake work, introducing the scene of Black alterlife, where it would be possible both to recompose history to transcend White supremacist forgetting while also strengthening Black resilience to weathering on the streets of Minneapolis and elsewhere. "It is to the breath that I want to turn now. To the necessity of breath, to breathing space, to the breathtaking spaces in the wake in which we live; and to the ways we respond, 'with wonder and admiration, you are still alive, like hydrogen, like oxygen.'"[66]

Sharpe moves the line between dreams of opportunity and threats of racial suffocation in her writing because, as I have noted in this essay, residence time splits somewhere between the ecological calculation for the time it takes to dissolve into vastness and to engender a Blackness that refuses a White dissolution of a body at rest. There are then metaphorical lines to cross in how scholars and residents imagine ecological life and death and empirical lines to cross as well. Deep water provides an ossuary, figuratively speaking, but also as a molecular exchange, and the air that we breath functions in much the same way. Anyone can track on a cell phone an Air Quality Index for their region, which is based (if the science is reported accurately) on the grain size of airborne pollutants and their accumulation over time. Particulate matter can be sorted by course, fine, and ultrafine thickness; PM2.5 is the agreed-upon metric for tracking acceptable levels of exposure to fine particles that someone inhales less than 15 μg/m^3 (micrograms per cubic meter) and that is how my dream map was substantiated at NOAA. The thicker, denser particles that the body labors to breathe still cling to the lungs of the contaminated survivors of 9/11. They are calibrated for "everyone with lungs during all the moments" through and for the very systems of scientific calculation purportedly invented to protect the living. In effect, though, the thicker, denser particles enacting the necroviolence of further killing. These lines that are drawn can and must be moved.

The Mètis scientist, Max Liboiron, knows firsthand through their laboratory work and then through their tribal heritage, regional respect for "Land" as an intimate, combined, and spiritual relationship to water and air. "Assimilated capacity" is the colonial gift that threatens to maim, sicken, or kill most living things by arbitrarily establishing thresholds for irretrievable pollution, a "sag" point on a statical rendering of when an ecological system can mend itself.[BM] Assimilated capacities were invented to universalize the belief that ecological systems can purify themselves, shutting the door so to say on residence time because the chemists, Streeter and Phelps, determined the point "when the oxygen demand of metabolizing waste exceeded the reintroduction of oxygen into the stream," and then over time, Ohio water was thought to be the same for all water, a colonial move to universalize all life through the powers of statistical reasoning.[CI] A simple, numerical truth metastasized across the twentieth century to underwrite different kinds of necroviolence, different "permission-to-pollute" systems. To end this essay, I propose these systems make it much easier for a society to tolerate pluri-violence by misunderstanding the arbitrary limits imposed by necroviolence, including how bodies and masses gracefully decompose only to be reborn again.[67] The same line through water, and through air, divides an endangered ecological world into permissible amounts of death and decay, when the lines we might dream about are the

BM Your earlier point about residence time made me think about Liboiron's argument about assimilative capacity and our reading group discussion of *Pollution Is Colonialism* (Durham, NC: Duke University Press, 2021). There was a lot I learned from this book when we read it together and, in particular, from their anticolonial methods and writing style, but the way you shaped how we worked through their argument about assimilative capacity helped that concept stick for me. In our discussion that day, you homed in on how racialized logics are constituted through the colonial assumption that the earth, rivers, air, or any ecosystem can absorb a certain amount of pollution before crossing a threshold of harm. This part of our discussion strengthened my ability to notice these types of logics and how racialization emerges from colonial structures like pollution (de)regulation. It was one of those precious (not rare but significant) moments in our reading group where I felt changed by our conversation, and for which I'm grateful.

CI The idea of assimilative capacity is so hideous. It's like saying you can beat your child to the brink of death, but as long as you stop before murder, so they can eventually recover, well then, everything's okay.

lines "between pollution and nonpollution . . . where policy, accountability and responsibility come together."[68]

I end with radical optimism folded into the embrace of an alterlife one year after tragedy in my community: there are lines everywhere that can be moved and are already moving.[CR] Antiracist scholarship and poetics help to move the lines of scientific and governmental permissibility, as long as stories of afterlife and alterlife are read with the seriousness of an IPCC report and with enduring knowledge that incident reports import a line that someone else has drawn, someone in a hurry. If there are lines that demarcate accessible levels of water and air pollution and they themselves are moving, two abolitionist projects appear. The first is to seek the wisdom of BIPOC accounts of violence in daily living because they know weathering better than others. Colonial science will not of its own accord translate residence time and assimilated capacity into new metrics for what is and is not a contagion, a pollutant, a threshold for acceptable social death. Empirical measurements such as these must be pulled from their colonial, supremacist roots and recast to invite antiracist engagements. Then we also need different stories, as phantasmagrams, that pin one dream to an abject system of death while sampling the same evidential stream to compose cultural registers of inclusive resiliency.[NS] Black, Indigenous, and Queer science alongside Black, Indigenous, and Queer stories that seek to "dream and feel other futures."[69] I have tried to share my own grief and need to tell a different story, aided by BIPOC atmospheric wisdom, by diving into the particulate world of inevitable death and dissolution. After the fires, the protests, the mass murders, the insurrection where I live, I need the lines to move because of the

CR I'd mentioned to you that reading your chapter had a powerful impact on me emotionally. You apologized for "bumming me out" in your joking/not joking way. Your work here is *hard* but not a bummer. It's more like you're noticing things I couldn't find language for exactly but already knew, or sensed, maybe, and somehow through these stories you tell, the words bypassed my thinking self and went straight to the heart of the problem and my heart. Dunno. If anything, it's because I recognized the abyss you speak of and felt it like a jolt when reading. This is also to say I felt less alone in that place of recognition.

NS I feel phantasmagrams, as you discuss them, are a vision of that rhetoricity I yearned for earlier in your essay—something that is not of the canon or tradition but that is about being present and in the presence of collective pain and death. I appreciate the move to let the exposure inform invention, recognizing the decomposition that infuses dwelling. I do wonder at stories, though. What if we don't need different stories so much as different ways of breathing the particulate?

sheer weight and magnitude of planetary collapse and injustice. I remember those lost to violence in thin relation now like a drop of rain, a misty breath. I taste the thickened dust from rogue prairie fire. One can reside in the alterlife if the burden of denial can be lifted.

Inconclusion **A Reading Group Meeting on *Rhetorical Climatology***

Candice I'm curious. How do you all want to start this conversation? Chris, you sent some questions. Should we start there?

Chris I don't have any way I want to "run" this or something. I think it can be organic. Those questions earlier were just provocations or prompts to get something started. One of the things we discussed last time was, now that we've all read the book we just wrote, we could treat this conversation almost like another book club meeting, where we're talking about the book as if from outside it. Of course, we can't be outside it; no one's ever outside what interests them, let alone from the structures or climates that partially determine what can be done in them. But I think that's also part of the book's point. Candice, why don't you share some of the things you wanted to say.

Candice Okay. Well, I was thinking about your question of what ties together all the different books that we've read, and the conversations we've had, and our writing. It feels like one thing threaded throughout are questions of social justice, ethics, and action within climates, given the complexities and complications. What is possible? How can we transform these hideously violent, toxic, racist, ableist systems?

Jennifer I would just interject in response to the questions you're raising, Candice, that we've had these ongoing side conversations that chronicle what's still missing or absent from rhetorical scholarship on these topics, which has been hugely formative for my thinking. And I think thinking with these absences through the lens of climate is particularly useful in how it gets to thinking about what totalizing impulse runs underneath and sustains them.

Candice Yes! Totally, Jennifer. And many of us seem to be foregrounding a kind of hope. Or maybe not hope, not a transcendent hope, but maybe more of an ethical orientation toward not only understanding but transforming conditions. "What might one do?"—those kinds of questions that keep turning back on themselves in these climate conversations, because everything is so complicated. And then this other thread of the impossibility of those questions of change, of transformation. Those two threads I saw throughout. How do you understand and navigate all-encompassing violence, whatever the violence might be? And what are the possibilities for a world otherwise? How might we act toward them?

Nathan I was thinking about that, too, actually. Not quite the way you said it, but that tension. I was feeling it reading it through again this morning, and I was reminded of a quote that my partner likes. When she was still teaching she taught Utopian studies, and she loves Calvino's *Invisible Cities*. And so this quote is what she mentioned:

> The inferno of the living is not something that will be; if there is one, it is what is already here, the inferno where we live every day, that we form by being together. There are two ways to escape suffering it. The first is easy for many: accept the inferno and become such a part of it that you can no longer see it. The second is risky and demands constant vigilance and apprehension: seek and learn to recognize who and what, in the midst of the inferno, are not inferno, then make them endure, give them space.[1]

And I thought that a lot of the different people we're quoting, and things that we're saying in the book actually sort of exemplify that way of thinking. That's hopeful, but it's sort of an ethical orientation. It's not "an ethic" in the sense of having a kind of specific framework for doing that work, and I was wondering about the extent to which, unexpectedly, thinking about climate forces us to confront the tension between pessimism and optimism that so many people are working between and in. And I think some of the readings we've done in the group really are dwelling in that space too.

John Nathan, I love that, and what I hear in that. And this is a way to go back to speaking with you, Candice. To me, that Calvino quote takes the Murphy article on alterlife, and gives it a kind of formalism and dimensionality. I'm not criticizing anybody. I'm certainly not criticizing Murphy, but that tension is really key to what I'm trying to do. I'm writing from a point of grief. I'm not making a policy argument *whatsoever*. There might be one down the road, but it's about grief and mourning and loss through violence. That then settles into that pessimism/optimism, which is an ideational space. But for Murphy, for the "residence time" that Christina Sharpe writes into—and that's *not* a metaphor; it's a powerful, biological calculation—that tension is kind of at the heart of what I'm trying to dig into, via the readings.

Jennifer The way you phrase this is really clarifying, the idea of fronting the emotional vantage point of our writing, not "flipping the script" *just* to highlight emotion but in order to crack open the door on the affective range that certain climates enable. I think writing about ableism at this point in time requires a sort of dogged anger that never fully tips into fury because the awareness is still so nascent.

Bridie I find this helpful too. One of the things I was noticing in reading your chapter, John, is how the theme of belonging comes up, which I think we each speak to in different ways but haven't named yet, at least not that I'm remembering. At one point in your essay you describe Murphy's alterlives and this idea of residence as a "belonging to the remainder." I know for me I draw meaning from belonging, from feeling like I belong to something, whether it's a team, organization, the department, whatever. But the histories that come along with belonging, whether we know them or not, seem to press in.

John You know, I'm just trying to work through my pain on behalf of so many other people well outside of my community, but also in my community, that have directly suffered. So, I'm just in a really kind of weird place. But it's also really exciting, in the sense that it feels responsible to try to say something into that. Long story short, I love the Calvino. I hear Murphy in that, and it links to Liboiron's "assimilated capacity," and that's kind of the argument I'm building. But then you ask, "Well, why?" I'm not trying to convince first responders to act differently, or for the meteorologists to get new instruments, or for campus and community to fund recovery work differently. I'm nowhere near that kind of, "Let's be rhetorical! We've got evidence! We could do problem analysis! We could propose a solution!" I'm just trying to get through the phenomenon of the grief.

Candice I hear you describing how we sit within violence, John, how it sits within our climates, and the many, many things that violence does to condition our collective lives, and does to our bodies, from shortening our telomeres to more profound, very overt violence. When I was reading through your essay, Nate, I found something similar resonating with me when you talked about trophism and relational worldmaking. You wrote, "the logic of nutrition that informs trophism deflects from the fact that trophism is inherently violent (what feeds on what and under what conditions)." You've mentioned things along this line in our group. And this influenced me to think about how social justice questions keep unraveling from this perspective. Yet, I keep wanting to come back. On what and under what do these human questions about justice or equity rely? The idea of trophism doesn't erase that problem, but it's a different axis you have to think through.

Nathan Yeah, I think so. I like the way you put that. I think it's really nice. It's been sort of my constant anxiety about the way that justice work, and equity and anti-oppressive work gets discussed in the field, because it's often (yeah, just sometimes, often is unfair and too broad) trying to recreate a space of heroism and purity in response. And I'm not a sad sack, but to some extent, those ethical systems that recognize to live is to actually harm others leave the question, "How do you live responsibly doing that?!" It's not an exciting way to think, but it feels real to me, in a way, every time I go into the classroom. I tell my students this: "You know that I see education is ethical violence." I have to make them different. I have to shove them, have to make them uncomfortable at times, or they're not going to be confronted with things they don't want. That's inherently antagonistic at some level. Even however nicely you do it, you can't disturb people's conventions without doing something that could be threatening to them.

And so the question to me is, "How do you do that responsibly?" But you're not going to live without doing harm. Of course, it's not spread evenly *at all*. And that's the equity issue. I really like the way you put that. Too often I think social justice questions, if they're grounded in the sense of *ending* harm or seeing violence as not enabling of their very efforts, they're gonna keep unraveling, and I don't know what to do about that.

Candice I don't either. I think the readings that we've been doing keep coming to this question, at least for me. I keep coming back to the same question over and over, "What does a social justice question look like from a climate perspective?" Nate, you end by saying that "forming new ecologies means forming new ways of inhabiting relationships with others in the fullest sense." For you, what happens to questions of social justice, of ethical relations, of equity? Maybe those words aren't even the right ones—but *less harm*.

Nathan Thanks for asking. If I really could fix it or have an answer, I wouldn't be sitting here, I guess. I don't know. What I was thinking when I wrote that, or when I've written things like that in the past, is part of my tendency to push against part of the public address tradition of rhetoric that I was trained in, which sees advocates as sort of world-changing figures. I appreciate that. They can be. They are on occasion. But it always sort of cuts out the frame of the picture in a way that leaves all that other stuff undiscussed. For me, to use Bridie's terminology, it's about changing those rhythms and the recursive dynamics of rhetoric. It's about changing those pulses and those rhythms by which we are present and going forward, and in doing so we're not going to actually stop climate change. We're not going to end the murdering of people of color. We're not going to get those complete absolute things, but what we can do gradually is to make things better. And what is better is its own set of problems, depending what we see there.

Jennifer This resonates with me and my motivations for thinking climatically. I don't think my writing is going to stop ableism, but hopefully there are one or two moments that might help a reader understand their own body differently and therefore orient themselves toward the world in a way that is not just less violent but more tender.

Candice I love this, Jennifer.

Nathan I like that too, Jen, that is more concrete and focused on weighing responsibility, and it also makes me want to add: I'm not a gradualist. Things can actually happen very quickly. But, for me, the way that things move is that the conditions under which they continue have to change. The specific actors and actions won't stick—not in a structural sense, but more in that tidal sense of rhythms that Bridie talks about: the kind of indeterminate rhythms and relations which have form but are not formal, fully.

But I feel that kind of work is super hard, because it's *so* dynamic. But that feels more like my actual life. I imagine with teaching pedagogy it feels like that: you have these systems of practices in place, but then each semester happens and it's kind of different. Things happen that don't quite look right or weren't anticipated. So, I guess that's a fuzzy way of answering, but for me worldbuilding means, top to bottom, rearranging everything, and having the courage to say to other rhetoricians, "Maybe the human really isn't the figure that we need to be thinking about. Maybe we should really just stop that for a while as a way of conceptualizing what we're doing." What would we get? Just bigger kinds of questions in that regard? Let me toss it to Bridie, then. How do you see tidal ethics as a way of doing that kind of moving and worldbuilding in one direction or another?

Bridie Thanks, Nate. I'm appreciating being able to listen to this conversation and how we're making sense of the books we've read together and the chapters we're writing from this reading, thinking, feeling, as Chris calls it. For me, I see this kind of moving and worldbuilding as really situated. I think in terms of the rhythms that have shaped my life and field work, and for me this means tides but an orientation to tides to guide how we might sense responsibilities and worldmaking likely isn't going to work in places that are not shaped by the forces and rhythms of tides. I mean, you could make an argument that tides are felt everywhere, but my point is not about trying to generalize this approach to other places but more about identifying ways to sense ethics, place, and relations together. There's diversity and unevenness in how we approach this across our chapters, but I hear all of us trying to situate ourselves in terms of these kinds of questions of ethics. And we're working through them in ways that are different from the typical ways we might position ourselves as researchers, or describe how our identities have shaped our work, or define something like how we gain access to a community. As we're working through grief or grappling with incommensurabilities and the complicities that we feel, we're asking "Am I really making things better, or am I just replicating the same kinds of forces and systems of oppression?"

I encounter this question and the need to consider rhythms, affects, and complicities in engaged work all the time. In terms of the rhythm, there are tidal forces that matter, and I've also found the breath to be really important as a kind of attunement, continually coming back to the breath. Michelle Murphy reminds us about this too, as she asks us to stay with alterlives. At a practical level, for me, it means meditating every day: breathing, so I can keep showing up, dealing with the difference, and pausing to let learning happen in my body, and to be sustained by that. I think about breathing as a practice of remembering, remembering to notice these kinds of forces and tensions, and how they play out, not just in my own body, but in many different types of bodies. Then, also, remembering the limits of that kind of embodied thinking, because it so easily slips into ableism, as Jennifer reminds us. So, yeah, connecting with tides and thinking with ethics as a way of staying with these kinds of proximate rhythms that nurture us but also that pollute us too.

Chris Breath is such a lovely way to take this for so many reasons, given the sort of airiness of climates. When yogis tell us to find our breath, they're not pointing us toward something that's not already there. It's an act of attention. In a lot of ways, these chapters are thinking about the kinds of attention we pay to the worlds we're in. And there's a logic, or a figure of thought that says, let's "think outside the box." But thinking outside the box is easy. It's thinking the place you occupy that's hard. How do we see where we stand, what we're doing in the box? How do we see inside the inferno without just pretending it's not there, or rather maybe creating a space for us to operate in a more humane way, or do less harm? And I think in some ways, like in the grief of John's chapter, there are some kinds of research where the "So what?" question just isn't a valid one to ask. That's not the kind of work it's doing, that many of us are doing. We're more interested in being in a space.

And the *being-in* is what something like antiracism is about. It's not saying we're going to topple this structure and create a whole new world. It's not a revolutionary act. It's daily work that we have to continue to sustain and practice.

Jennifer Fully agree. And for ableism, it's really daily work on one's perception and inhabitation of one's own self. We are all only temporarily able-bodied, as the quote goes, but living that knowledge is a way that isn't fear-forward requires a great deal of discipline and patience.

Chris Candice, I know you've done a lot of antiracist work, and you talk about it in your chapter through pedagogy and teacher development, particularly. We're doing some of that work in our department, too, and the conversation keeps coming back to it not being a box you check and clear off your list. It's something that you have to do over and over again. It's not like you put a policy in place and then everyone's all set because now we can punish somebody if they do something awful. It's, no, we have to continue to do this over and over. We can't ever get outside this box. We have to think the place we occupy, and that's the way I come at the worldbuilding when it comes to doing less harm.

Bridie Yes, and the way you are describing this makes me think about Candice's chapter and the conversations we've had over the last few years about the kind of change making we're trying to do on our home campuses and communities. Worldmaking is repetition, and it is sometimes hard to know what kind of difference it is making.

Candice In so many of our readings and conversations, we talk about climate as "all-encompassing," "atmospheric," as "particulates we breathe in." Bridie, in your work, I could hear the need for care in the face of this; breath and meditation and care as a way to stay with work. You need a particular kind of disposition to imagine change from a climate perspective. Because we are immersed, there is no escape, and yet we are trying to find a way forward, or just a different way.

Bridie In one of our earlier sessions, Chris, you asked the question of what we see as the difference between ecology and climatology, or, you know, this focus on climates. I struggled with that question because I don't know that I really see a distinct difference. I think there are maybe discursive differences in how those concepts come to make sense. But, Candice, what you're saying, it does seem like this climatic thinking provokes a sense of—it's not so much enmeshment, which feels like more of an ecological orientation—it's more like *entrapment*. There's an inescapable quality to the climates that we inhabit. This has consequences for agency, of course. And I don't know what agency becomes within climates, but maybe it's like our earlier conversation about that recognition of capacity and ability always being in intimate relations with each other, where it's debilitating to recognize that we really have no control. But we keep trying.

Nathan I wonder if it has to do with what we want to see our actions accomplish. It's one thing to talk about the kind of care we're discussing here, versus the immediacy of specific actions which can relieve particular suffering, or provide some greater sense of fairness and justice and equity, as you were discussing, Candice. In the face of those sorts of things, if it gets to a level of crisis, it produces a kind of forgetting that actually can be counterproductive. But, on the other hand, that slower, more patient sort of morbid sense of what's possible, these don't seem like very good things to focus on. Yet, when I think about broad change over time, which can and does happen—you know there's *nothing but* constant change—so that should give people hope. That does happen in very strange ways. To me, again, it goes back to the sense of not just the human, but the individual in rhetoric, of wanting to find the mode of control in a specific form of action, with measurable consequences relative to those actions, which may not be a really workable way to imagine making the world different. And at that point the kind of focus on self, even in the sense of caring for oneself, re-creates that tension of the particular and the general.

I feel sometimes, when people are engaging in this kind of social justice work, they fall into the habit of acting as if, if they don't do it the right way, or if they do do it the right way, that's the whole point. Their own self-performance *is* the outcome. To me, from my own subject position, I'm a permanent failure, and will be a permanent failure at the things I care most about. Because I *have* to be. It couldn't be any other way. And so, coming to terms with the fact that I am absolutely going to fail at the stuff I care most about, but the fact that I keep trying is the whole point. I keep trying to fail because I will fail, but not trying is absolutely not acceptable. And I find hope in that. Not that I'm going to suddenly become this magical person that I wish I was, or that my actions would have the kind of outcome that I wish they would. And that to me is a different orientation to what exactly can I do. And yeah, it sits in that tension that we've been talking about.

Candice This reminds me of Christina Sharpe's work, among other Black feminists who think about worldbuilding. The act of imagining a world otherwise *is* a praxis. That's asking very different questions than the ones that you just brought up, which are like, *Am I getting this right? Do we have the social justice plan in place? Then let us execute it, and now we're good. If we've failed, that's because the plan's not good*. Instead of this overemphasis on "controlling" and a "plan," Sharpe's work is asking something different.

Nathan One of the paradoxes of that sense of control that you just mentioned, where the antiracist work, anti-oppressive work, the anti-anti-queer work, all those sorts of things come together, and especially in the sort of accounting for privileges, which is a very important dialogue to have. But it actually redoubles Whiteness in the sense that you're supposed to transcend your condition, and supersede it. So it's saying, particularly to people enacting forms of Whitened being, that they need to be *more* White to take *more* control, so that they can self-transcend their circumstances, and not be that way. And a different ethic, which is more in-line with what you're saying about Black feminism, is to recognize the way you're embedded and will not actually do that. But it's not about whether you're being a good or a bad person, but rather the work you carry to sort of go forward within that. That's a very different sense than the one which is like, "Oh, I'll gain control," which is just reactivating a sort of *homo modernus* in Silva's sense. Well, maybe that's not actually the way to go about it.

John Candice, I so respect and love you that hearing that my chapter is a bummer—

Candice No, it's not a bummer! That's not the message.

John I know, but it makes me want to rewrite it so that I can end on a really happy note. I'm just kidding, I'm just kidding. But the inescapable has come up in this conversation three or four times as a middle space, a gap between optimism and pessimism, a different kind of imaginary. There has to be some practical complicity living in the middle of an ecological world, which is empirical in some ways, but which is metaphysical and spiritual, and thus grief-receptive in another way. I don't quite have the language for what I want to talk about. But that is maybe the only positionality that is not rhetorical, which is not pragmatic, which is not all that modernist and very White posturing of "We can struggle through the problem together and come up with a better tomorrow." I'm hearing lots of comments on that. Just as an example of the possible hope, understanding in what ways grief, and commonality of grief, and the dissolution of the body into the molecular happens. The alterlife that is molecular is super powerful for me. I grew up in a Protestant household with the whole ashes-to-ashes, dust-to-dust stuff. That's not what Indigenous thinkers and scientists are talking about. They're talking about an entirely different notion of ash.

I'm thinking about necroviolence, which is a play on the Necropolitical by Jason de León. Necroviolence is those insidious twists of raw systemic violence that turn the ecology—it could be the sea, could be the desert, could be the city—into a tool of destruction. And the hopeful part—no, "hopeful" is the wrong word—it's just a fresher, more honest kind of realism about the lords and tools of war and destruction, which are massive in our times, right now. And if we know that this is necroviolence, we actually *look* at the normal ecological phenomenon that we see, and maybe take for granted. I don't think *you all* take them for granted; I think one benefit of the group is that we're more sensitive to how rich and uncharted a term like "ecology" is right now, or "environment" or "situation." But the hopeful side of necroviolence is that you actually see the full scope of the mechanisms of war. And if you see that, then there's an ethical opening—I think, I hope—that maybe you talk to somebody differently, or you teach differently, or you look at policy differently, or you put your nose in the wind and you smell and you breathe differently, and it's more capacious, and it's more loving. There's a kind of different positionality that I'm hearing people talking into today, which is really super serious. But it is the best we can do in the circumstances. I'll speak from my body in my position: White folks ain't taking us here. The pathway is not twentieth-century White pragmatism. We're heavily dependent on Black, Brown, Asian, Indigenous thinkers, because they already live there. That's been their world for the longest of time. That was a long comment, and I apologize.

Nathan It was great, though.

Candice John, your work is hard to read, but it's not a bummer. It's something else. I feel like all of us are asking in different ways, "What does a climate perspective or sensibility allow us or force us to pay attention to?" And you're asking us to pay attention to different kinds of violence, and where violence goes, where it sinks, how it enters the body. You don't overcome violence; it is atmospheric and sustained, even if it changes forms. Evidence of violence remains. We are all working together in ways we cannot comprehend to condition the shape of the violence that shapes us in our climates. Again, like a broken record, I keep wondering, how do the insights you share change the aims and shape of social justice questions?

Nathan I really like that last exchange between you two a lot. Actually, I think that's really getting to the heart of a lot of issues across the whole book project. I think the hope actually lies *in* the violence, and I think that's what Sharpe is saying, for example: that, in fact, we have to appreciate that all this pain and suffering, all the traumas, are actually generative. But that's not a redemptive thing. That's not a way of excusing it. In fact, it's taking very seriously that pain is not for naught; it's not pure negativity. It does produce creativity. And if you can't find hope there, you probably won't find it at all, because there's so much violence that you've got to be able to appreciate the depth of that. Try not to be overwhelmed by it, and then figure out, "How do I create out of that?" How do I find joyfulness with recognition that the world is really in a very bad place? It has been for a long time, but seems to actually be finding ways to get worse. And that compels us to find a different ethical orientation that does not assume a kind of triumphalism or a simplistic *overcoming*.

Bridie That really resonates with me, and I think it does across all of our writing where we're grappling with these tensions, the questions Nate's raising, in different ways. And it makes me wonder what does it mean to seek hope within violence? What's that process? I really appreciate how Tiffany Lethabo King describes her shoaling concept as a methodology of listening, but listening in a way that's very different from the dominant cultural understanding of listening, which is so passive and non-verbal. She's interested in listening that's rooted in humility, a kind of open vulnerability to let the world in, to let ourselves be changed in that process.

This may seem like a bit of a jump, but you know, I think that one of the unique things for me out of this academic space that we've been in together is that it *is* a space of listening, listening to each other, and then listening to these diverse voices we're reading. Scholars who I haven't read before, which changes who I listen to and cite in my work, and I'm changed through that too. Working in other academic spaces, you know, the day-to-day, the mundane, going to faculty meetings and just having all these task-oriented things, or even going to a conference: these kinds of interactions can be so constrained. This space affords a different kind of listening practice with each other. It's not a "best practice," or any kind of recommendation, but I do think there's something generative here.

Chris I think one nice thing about the group, too, has been that we've been at it for so long that, even though all our conversations eventually come to an end, we know there's going to be more conversation down the road, so there never seems like a final word. The door is still open. You know there's no tidy ribbon that we're trying to put around a book, say, or somebody's comments, because that's not true to the complexity of things. It would be convenient. It would be a lot nicer if we could say, "Oh, I got that figured out."

Nathan Just reacting to your comment there, Chris, it reminds me of another thing that you'd asked in one of the questions: What annoys us about the study of rhetoric? I don't want to go too far in that direction, personally, but one of the things I did think is more of a conceit that scholars will have, especially in fields that fancy themselves paying attention to politics, is that the writing and the work that we do in those sorts of spaces are actually in some way directly connected to initiating the changes we would like to see. And they're not, frequently, at all. They really operate in a completely other space. The engaged work is a way to try not to do that, but even there it's challenging.

And so, to me, the sense of what's the work of thinking through those problems, in forms like this or otherwise, recognize that it doesn't have any direct levers into the kinds of issues we're actually describing, which is not to say it's an excuse or wrong, or anything like that. I'm just saying that it's doing work, but maybe not the way it's framed. The typical one I hear is people say, "I'm going to make an intervention in . . ." And when I read it, I always say, "In what sense is that really being an intervention?" So, what is the work of the thinking, and the rethinking, and the changing, recognizing that it's very valuable for us to become different and better at what we do, but knowing that it actually doesn't directly plug into some of those problems. To me personally, it's about being responsible, but also recognizing my limits so I can figure out what actually is my contribution, or could be, and being really sober about the tininess, at least for my own life, of what that can mean. But, again, I find real hope in that sort of thing. I don't like the idea of imagining I'm some sort of person who makes things different by writing stuff. That would be a little bit disheartening to me.

Candice Bridie, I would be curious about what you think as someone who does such collaborative engaged work.

Bridie I've been interested in how much we've been talking about hope today, because I have a fraught relationship with hope as an emotion. There's an article that I assign in my classes just to kind of provoke and encourage students to respond to something, and they don't typically like this article. The title is "Beyond Hope," and it was written by Derrick Jensen and published in *Orion Magazine*. The opening lines are basically, and I'm paraphrasing here: Fuck hope. Hope is what keeps us chained to systems that are broken. And in some ways this resonates with me, honestly. I don't often find hope to be very motivating. I find a lot of the work that I do to be shaped more by grief and anxiety than by hope. So, it's like living with that kind of condition and trying to figure out how to keep showing up for that. But to Nate's point, to keep that showing up focused and take time to sense what the world is asking of me. I'm intuitively trying to track that, which is effortful, and it also requires that I let some things go, choosing some directions and not others, saying no. I also think about political commitments in research, like what Kim TallBear describes as "studying up." In your chapter, Candice, I think you're drawing from Ahmed's directional politics, which is a similar idea. It's never really clear if one knows what the direction is, or what "up" means, or who's in power at any particular moment, but it's kind of an orienting commitment to reflect on. I find it at least has some level of heuristic value, especially when we're in such a complex situation, as Chris was saying earlier.

Chris I think one thing that's going on, too, and not to go back to rhetoric specifically, but just in the academy in general there's an inculcated sense of how we're supposed to read and what our work is supposed to do. It's supposed to be citable; it's supposed to have a clear thesis, keywords that succinctly summarize everything you're trying to accomplish and say, you know, 150-word abstracts. That all lends to a kind of extractive way of approaching the text at hand, which in turn lends to a soundbite distillation of complexity, and the sort of desire to imagine that the solutions are simpler than they actually are. And that's the kind of violence that a strongly anthropocentric version of rhetoric—largely in the sort of public address tradition, I think—is complicit in enacting. One thing I've liked about all the books we've read, I'd say, is that they're very nonviolent books. They're very complex books. But I'm a reader who, if you ask me to tell you what a book is about, I might have a harder time telling you that than telling you what it felt like to spend time with it. And these books do that in a way that is way more, not just fun or enjoyable to spend time with, but it's more mind-altering for me to read some of these books, because they feel like they invite less of that extractive approach to them. And I think, without having talked about it, we're all kind of doing some of that work in our book, which I really appreciate about y'all and am grateful that you think that's important.

John I like that.

Nathan I do too. What you just said, Chris, it does encapsulate sort of what I've always wanted to do as a writer. I can see that as sort of an aspiration. I guess what I've had at heart was trying to write something that I think people would want to sit with, as opposed to something that would, you know, be quotable or something like that. I didn't think of it on those terms, but the way I've always thought about it is people who want to be published to build a brand, or to do that sort of thing. And again, in terms of the extractive piece, I think that approach to scholarship, which has many critiques of it—there's a ton, and not just the ones I'm making—goes in the opposite direction than being community-building. So, I think it's a really nice way to try and imagine what the writing task is.

Candice Your last comment, Bridie, gives me language and ways of thinking about how I think about hope from within a framework that also acknowledges that anything we do will be complicit, partial, something that can do harm. I was also thinking about the way positive narratives, or what we think is productive or healthy, or "the public good," can be where violence hides most insidiously. I think of your work, too, Jennifer, and how you highlight the heinous ableism rooted in narratives of "overcoming." And, Bridie, your turn to beloved images of Earth in environmental movements—or in my case looking at certain kinds of relationships to access in education that we think are positive—but actually these are the places where violence is most hidden.

Bridie Yeah, I really appreciate how you're characterizing that. And out of this project I feel like I have an increasing sense of the level of distrust that I need to have, and the things that I'm personally invested in or, to use Jennifer's term, inspired by. Like you're saying, there's the very real potential for hidden violence there.

Jennifer I think we're all talking about how certain violences are continually assumed to be inevitable. Not framed as such—that feels too intentional—but just taken as inevitabilities. And then the question is not only "what do we do with that?" but "what as rhetoricians specifically can we do with that?" I don't think erasing violence is a possible goal, but we are all probing for where we might be able to mediate it in meaningful ways.

Nathan I think so, too, but it's even more upsetting, I would say, and that's why I use the language of trophism in that particular part. Because for me to do anything, I must do harm. I have to be nourished by something else which is not going to continue. I really don't have a choice in that. No one does. John, have you read Elizabeth Povinelli's essay "The Child in the Broom Closet"? It's on Ursula Le Guin.

John Yes.

Nathan There's that wonderful way she pins you down, like the way she explains Le Guin, like, make that choice to walk away from Omelas. There's no future anterior; you're not imagining the world as it will have been. You have to make a decision *now*, based in what you can compromise on, based on the conditions that are. That's the ethical point that you're actually occupying. And for me, I think trophism is a way of thinking about this, a way to imagine that in a broader, systemic way. There is certain harm that I will and must do to continue to do things that I see as positive as well. And what forms of harm do I not want to continue? And how do I weaken those and make them go away? But I will in fact further other systems of harm because I can't help it. I can't actually do what I need to do if I don't. And that to me is really ethically complicated, but it's actually more accurate. And it forces people to be responsible for the harm that they will do, not because harm is intrinsically bad, but because it's *intrinsic to being*.

Candice Yeah. That resonates with me so much. You are highlighting the importance of the choices we make and how crucial it is to be accountable to those choices and pay attention to the impact of whatever harm is caused by them. It kind of eliminates the moral high ground that many asking justice questions often want to inhabit, well, maybe we all want to inhabit. It feels better that way. But, when I think about the pragmatics of putting these ideas into practice, it's a *hard* sell, and doesn't exactly always inspire groups to action: *Let's get cracking on our work toward justice* (*but don't forget to reflect on how we are messing up, complicit, and doing violence, too!*). *Go team!* I dunno. Clearly, I'm sitting in these fraught places in my work.

Chris During this conversation I've been thinking about a couple books that we didn't read as a group, but you may all know them. One is Jane Bennett's *The Enchantment of Modern Life*, and the other is Harney and Motens's, *The Undercommons*. In the *Undercommons*, they talk about hope being a form of oppression. There's some line in there about hope being the policy thrown like tear gas into the undercommons: this idea that, here, we're going to tell you hope is how change happens, because we have to create a space for you to give effort, for you to care enough to make an effort to do something. At the same time, that's not enough. The circumstances are dire enough that hope is not gonna fuel this car. So there's got to be something else. And I think in their more recent collaboration, *All Incomplete*, they emphasize a version of what we're talking about: the listening, the holding things open. They're not trying to say they've got this figured out because things are still ongoing, and are going to be, and if there's a space for hope there it's only provisional and with asterisks. And that's where the *attention* comes to matter. Which is how the Bennett book came to mind, I guess, in that part of her argument, as I remember it, is to say, Look, ethical systems fail when they just tell us which behaviors are right or wrong. Do this. Don't do that. We need to have, as Candice was saying, a disposition to hear this or a disposition to act in a certain way, and the way to achieve that is through attuning or attending to different things. Attention has come up a ton in this conversation. And, of course, there's Richard Lanham, who talks about rhetoric as the art of attention; so there's already a tradition within rhetorical studies of thinking about attention. We're all sort of pointing readers' eyes, I guess, to different ways of doing that.

Bridie This is making me rethink what I said earlier about the emotions or dispositions that shape my own engaged work, that it's motivated by grief and anxiety, because that's not a complete enough description. Candice, when you mentioned Saidiya Hartman's work, one of the things that I took away from her writing was its sense of intimacy. She uses the concept of eroticism, but not in a sexualized way. It's *like* love, but it's more than love in terms of the kind of attachments that come through these modes of attunement. And I think that's also really a central part of collaboration and knowledge coproduction. Feeling an intimate sense of connection with a place, with difference, with many different entities, people included, who I wouldn't otherwise encounter, and life becomes a different kind of thing through that. I mean through her narrative, which prepares us to receive the context that she's describing.

Nathan That was a wonderful comment, by the way. I love that book for the same reasons: it's the intimacy she creates. She's a magical writer. I mean, honestly, I would kill to be that good with words. I actually wouldn't kill, but I would like it if I was that good with words.

It's strange that this conversation has become about ethos, not in the strategic, deployed sense, for persuasion, but ethos in the sense of a kind of felt way of being, and how that affects other things around you. So, again, a more climatological view of ethos is where it seems like we've gotten to, and that's maybe going back to what you say right off the top there, Candice, about a way of imagining how do you mediate between that individual or the specific and the general.

There's a place to imagine *ethos* in a more climatological way that tries to balance them. There's a sense of being situated, not of hopelessness, but a non-transcendent situatedness, that is also affective. We can see it as somehow pushing forward toward something better, without a hope that the better will actually show up on the doorstep tomorrow, or even within my lifetime. I'm always reminded of when Karlyn Campbell would teach the early women's rights rhetoric, and she's such a powerful teacher. She found these great lessons and things you wouldn't pay attention to, and she would always say, "Cady Stanton never saw the right to vote, and yet she spent her entire life fighting for it." What do you think about that? Just giving yourself over to something without ever seeing it come to fruition? I'm sure she felt disappointed. But the fact that she would do that; that's *ethos*. It's not about the strategic use of *ethos*. But it's *ethos*. So maybe that's an unexpected rhetorical direction for thinking climatically.

John These are beautiful statements. They really are. But I have to go attend to my people.

Nathan Before we go, this conversation is making me think of the book's title, *Rhetorical Climatology*. A lot of what we've been saying would make that title kind of an interesting point, like, *Hmm . . . What do we mean by that*? It's just interesting to me.

Chris I can share my thinking on it. The idea of "climates" sort of captures the right scale for the project, anyway it was the word that I settled on among many others to think about a surround, as Murphy might talk about it. But I didn't like "rhetorical climates" because that felt like a whole bunch of discrete boxes, whereas "rhetorical climatology" might suggest a praxis or a study of something, a way of doing something. How does Liboiron talk about methods? As a way of being in the world? That's sort of the register that I hope the title hits. As this conversation underscores, I think we're all thinking in a similar way, applying our approaches in our own unique perspectives on the work we're doing in our respective contributions. But I felt like that scale, and that tacit emphasis on practice or ways of being might do some of that work. Or maybe you all don't like it.

Nathan It's not like or not like. It's always the curiosity. It's a question. So it does something, and there's no way to put those words next to each other without making one of them modify the other in some direction. Whatever you do is just interesting, so I appreciate that.

Bridie I think some of the work that you do, Chris, in your chapter, invites the problematization of climates through an immediately critical orientation. That sets up the work that we do in the book so it's not saying that we're offering this term as an answer or solution. It's saying that this term has a history tied in with determinism as a racist concept. I think it's a productive way of opening up the book, starting with a digression and criticism. I'd be less comfortable using the concept if we weren't doing that work in the opening chapter, because that does invite an orientation for understanding, and I think it models that sense of what we've been talking about: how to find a way to work with the harm.

Nathan But there's no way we're going to talk about anything without activating some structures we don't really like. So better do something, you know, seemingly constructive with those. Thanks, Bridie.

Bridie It seems like we've talked a lot about a lot. But one other thing I wanted to raise that I think we're talking about is scale. When we're talking about scale as a surround or as all encompassing, or as an atmosphere, I wonder if we haven't talked about *time* enough specifically, as an aspect of scale. Or, how we can think in scalar ways. And, again, Chris, your opening chapter, in the way that you're talking about the difference between weather and climate as a difference of time, it makes me think about the role that time plays in our discussions about climates. You know, in tidal ethics, the tuning to tides is a tuning to a multiplicity of times.

Chris I had a whole thing on the temporality turn that I cut out of the introduction! But I do think time is key here.

Jennifer Yes, definitely. So many discussions around justice revolve around bettering the present, which is definitely important. But cracking open the question of who has temporal options, multiple futurities, is much harder.

Nate Yeah, I think it matters too. And, again, it's not like there's a right or wrong way. But what is it we mean when we're talking about time? You say there's a multiplicity of times, but also at the level of change we've talked about, I've often thought you can imagine time as contingent change, that *things might not be this way*, and so that implicitly implies a sense of temporality where things may or may not be. But then there's also a sense that things will inevitably *not* be this way, which also implies a sense of temporality. One is more sort of thrown, the other one is more contingent. Both of them toggle on the sense that there's always more than one way. And time to me is like an acknowledgment of that, but it takes account of it through some sense of duration or looping or continuation. So when you say there's multiple senses of time, to me that means there's multiple senses of dealing with the more-than-can-be-now, which to me is—not to say hopeful—but I find I get a lot of energy out of the simple fact that nothing really endures forever. That actually gives me hope, honestly. Anyway, those are some of my thoughts about that.

Bridie That would be a hopeful way to end the book.

Notes

INTRODUCTION. **WHAT WE TALK ABOUT WHEN WE TALK ABOUT RHETORIC**

1. Wayne C. Booth, *The Rhetoric of Rhetoric: The Quest for Effective Communication* (Malden, MA: Blackwell, 2004).
2. Nathan Stormer, "Rhetoric's Diverse Materiality: Polythetic Ontology and Genealogy," *Review of Communication* 16, no. 4 (2016): 302.
3. There are too many people working in this area to cite them all. For some overviews, see Lawrence Frey and Joshua Hanan, "Toward Social Justice Activism Critical Rhetoric Scholarship," *International Journal of Communication* 14 (2020): 850–69; Stephen Hartnett, "Communication, Social Justice, and Joyful Commitment," *Western Journal of Communication* 74, no. 1 (2010): 68–93; JongHwa Lee and Seth Kahn, eds., *Activism and Rhetoric* (New York: Routledge, 2020).
4. Gerard Hauser, "Teaching Rhetoric: Or Why Rhetoric Isn't Just Another Kind of Philosophy or Literary Criticism," *Rhetoric Society Quarterly* 34, no. 3 (2004): 39–53.
5. This research, too, exceeds what can be cited here, but for some touchstones, see Sara L. McKinnon, Robert Asen, Karma R. Chávez, and Robert Glenn Howard, eds., *Text + Field: Innovations in Rhetorical Method* (University Park: Penn State University Press, 2016); Michael Middleton, Aaron Hess, Danielle Endres, and Samantha Senda-Cook, *Participatory Critical Rhetoric: Theoretical and Methodological Foundations for Studying Rhetoric in Situ* (Lanham, MD: Lexington Books, 2015); Candice Rai and Caroline Gottschalk, eds., *Field Rhetoric: Ethnography, Ecology,*

and Engagement in the Places of Persuasion (Tuscaloosa: University of Alabama Press, 2018).

6. Michael McGee made an influential variation on this point back in 1990. See Michael McGee, "Text, Context, and the Fragmentation of Contemporary Culture," *Western Journal of Communication* 54 (1990): 274–89.
7. For more on this notion, see Tim Ingold, "That's Enough about Ethnography!," *HAU: Journal of Ethnographic Theory* 4, no.1 (2014): 383–95.
8. Audre Lorde, "The Master's Tools Will Never Dismantle the Master's House," in *Sister Outsider: Essays and Speeches* (Berkeley, CA: Crossing Press, 1984), 110–13.
9. Astrid Schwarz and Kurt Jax, "Etymology and Original Sources of the Term 'Ecology,'" in *Ecology Revisited: Reflecting on Concepts, Advancing Science*, ed. Astrid Schwartz and Kurt Jax (New York: Springer, 2011), 144–47.
10. Hannah Arendt, *The Human Condition* (Chicago: University of Chicago Press, 1958).
11. Schwarz and Jax, "Etymology," 147.
12. Arendt, *Human Condition*, 200.
13. Timothy Morton, *The Ecological Thought* (Cambridge, MA: Harvard University Press, 2010).
14. Karen Barad, *Meeting the Universe Halfway* (Durham, NC: Duke University Press, 2007); Sofie Sauzet, "Phenomena—Agential Realism," *COST Action IS1307 New Materialism: Networking European Scholarship on 'How Matter Comes to Matter'*, March 13, 2018.
15. Jenny Edbauer, "Unframing Models of Public Distribution: From Rhetorical Situation to Rhetorical Ecologies," *Rhetoric Society Quarterly* 35, no. 4 (2005): 20.
16. Nathan Stormer and Bridie McGreavy, "Thinking Ecologically about Rhetoric's Ontology: Capacity, Vulnerability, and Resilience," *Philosophy and Rhetoric* 50, no. 1 (2017): 2.
17. Dobrin's CV is prolific, but for some standout monographs on ecological rhetoric in the context of writing, see Sid Dobrin, *Postcomposition* (Carbondale: Southern Illinois University Press, 2011); and *Natural Discourse: Toward Ecocomposition* (Albany: State University of New York Press, 2002).
18. Joshua Trey Barnett, "Rhetoric for Earthly Coexistence: Imagining an Ecocentric Rhetoric," *Rhetoric & Public Affairs* 24, nos. 1–2 (2021): 365–78.
19. Thomas Rickert, *Ambient Rhetoric: The Attunements of Rhetorical Being* (Pittsburgh: University of Pittsburgh Press, 2013); Diane Davis, *Inessential Solidarity: Rhetoric and Foreigner Relations* (Pittsburgh: University of Pittsburgh Press, 2010); Megan Eatman, *Ecologies of Harm* (Columbus: Ohio State University Press, 2020); Debra

Hawhee, *Rhetoric in Tooth and Claw: Animals, Language, Sensation* (Chicago: University of Chicago Press, 2017).

20. If you're curious how the life of different existents is rhetorically assigned a value, have I got the book for you: Allison Rowland, *Zoetropes and the Politics of Humanhood* (Columbus: Ohio State University Press, 2020).
21. For an argument about "working through" being the work that rhetoric does, see Jens Kjeldsen, "Rhetoric as Working Through" (paper presented at the Norwegian Media Researchers Conference in Bergen, Norway, October 20–21, 2016).
22. This notion is inspired by the lovely phrase, "futurity without optimism," from Allyse Knox-Russell, "Futurity without Optimism: Detaching from Anthropocentrism and Grieving Our Fathers in *Beasts of the Southern Wild*," in *Affective Ecocriticism: Emotion, Embodiment, Environment*, ed. Kyle Bladow and Jennifer Ladino (Lincoln: University of Nebraska Press, 2018), 213–33.
23. Look no further than our own Nate Stormer for an argument to this effect: Nathan Stormer, "Rhetoric's Diverse Materiality," *Review of Communication* 16, no. 4 (2016): 299–316.
24. Anna Lowenhaupt Tsing, *The Mushroom at the End of the World* (Princeton: Princeton University Press, 2015), 285.

CHAPTER 1. **DIGRESSION ON AIR**

1. For more on air, breath, and racial respiratory philosophy, see Matthew Houdek and Ursula J. Ore, "Cultivating Otherwise Worlds and Breathable Futures," *Rhetoric, Politics & Culture* 1, no. 1 (Summer 2021): 85–95. For work on air quality, climate, and respiration in the context of disability studies, see Alison Kenner, *Asthma Care in a Time of Climate Change* (Minneapolis: University of Minnesota Press, 2018). See also Catarina Albano, *Out of Breath: Vulnerability of Air in Contemporary Art* (Minneapolis: University of Minnesota Press, 2022).
2. Norton Juster, *The Phantom Tollbooth* (New York: Yearling Books, 1961), 101–8.
3. Plato, *Timaeus and Critias*, trans. Desmond Lee (New York: Penguin Classics, 2008), sections 16–21.
4. Sara Ahmed, "Not in the Mood," *New Formations* 82 (2014): 13–28.
5. NASA, "NASA—What's the Difference between Weather and Climate?," NASA February 1, 2005, https://www.nasa.gov.
6. "Thing-power" is Jane Bennett's provocative term, most developed in her book on

vital materialism: Jane Bennett, *Vibrant Matter* (Durham, NC: Duke University Press, 2010).

7. For more on polysemy in the context of rhetoric, see Leah Ceccarelli, "Polysemy: Multiple Meanings in Rhetorical Criticism," *Quarterly Journal of Speech* 84, no. 4 (1998): 395–415.
8. See, e.g., William Keith, Steve Fuller, Alan Gross, and Michael Leff, "Taking Up the Challenge: A Response to Simons," *Quarterly Journal of Speech* 85 (1999): 330–38; and Edward Schiappa, "Second Thoughts on the Critiques of Big Rhetoric," *Philosophy and Rhetoric* 34, no. 3 (2001): 260–74.
9. Nathan Stormer and Bridie McGreavy, "Thinking Ecologically about Rhetoric's Ontology: Capacity, Vulnerability, and Resilience," *Philosophy and Rhetoric* 50, no. 1 (2017): 1–25.
10. William James, *Pragmatism: A New Name for Some Old Ways of Thinking* (New York, 1907), 97.
11. Mike Hulme, "Climate," *Environmental Humanities* 6 (2015): 175.
12. Stephen Tyler and Ivo Strecker, "The Rhetoric Culture Project," in *Culture + Rhetoric*, ed. Stephen Tyler and Ivo Strecker (New York: Berghahn Books, 2009), 21.
13. Hippocrates, *Hippocrates on Airs, Water, and Places* (London: Wyman & Sons, 1881).
14. Aristotle, *Politics*, trans. C. D. C. Reeve (New York: Hackett Publishing, 1998), VII, 23–26.
15. Aristotle, *Politics*, VII, 26–28.
16. Aristotle, *Politics*, VII, 28–32.
17. Dan Buettner, *The Blue Zones: Lessons for Living Longer from the People Who've Lived the Longest* (Washington, DC: National Geographic Society, 2008).
18. For a helpful overview of the historical classification of climates, see Marie Sanderson, "The Classification of Climates from Pythagoras to Koeppen," *Bulletin of the American Meteorological Society* (1999): 669–73.
19. The dearth of translations available have left important Arabic work in geography largely unheeded in Western contexts, and I have had to rely on secondary sources. The foremost overview of the topic seems to be André Miquel's *La Géographie Humaine du Monde Musulman Jusqu'au Milieu du XI Siècle* (Paris: Mouton & Co., 1975). Also see Ralph W. Brauer, "Geography in the Medieval Muslim World: Seeking a Basis for Comparison of the Development of the Natural Sciences in Different Cultures," *Comparative Civilizations Review* 26, no. 5 (1992): 73–110.
20. One such book has been particularly helpful as an overview and incitement to more

primary sources. See Lucian Boia, *The Weather in the Imagination* (London: Reaktion, 2005).

21. Ibn Khaldûn, *The Muqaddimah: An Introduction to History*, trans. Franz Rosenthal (Princeton: Princeton University Press, 1967), 167–76.
22. Ibn Khaldûn, *The Muqaddimah*, 168–69.
23. Ibn Khaldûn, *The Muqaddimah*, 171.
24. As Joshua Trey Barnett points out, the belief that humans can't change the weather has a long history, though it is no longer true. In a time of anthropogenic changes to the atmosphere—as well as geoengineered proposals to ameliorate them—weather and the air itself are very much under human sway. Joshua Trey Barnett, *Mourning in the Anthropocene: Ecological Grief and Earthly Coexistence* (East Lansing: Michigan State University Press, 2022), 60.
25. Ibn Khaldûn, *The Muqaddimah*, 168.
26. Boia, *Weather in the Imagination*, 27.
27. Boia, *Weather in the Imagination*, 57.
28. Richard G. Barlow, "Infinite Worlds: Robert Burton's Cosmic Voyage," *Journal of the History of Ideas* 34, no. 2 (1973): 292.
29. Robert Burton, *The Anatomy of Melancholy*, The Ex-Classics Project, 2009, 43, https://www.exclassics.com.
30. Barlow, "Infinite Worlds."
31. Burton, *Anatomy of Melancholy*, 54.
32. Boia, *Weather in the Imagination*, 39.
33. For some highlights, see Rob Nixon, "The Anthropocene: The Promise and Pitfalls of an Epochal Idea," in *Future Remains: A Cabinet of Curiosities for the Anthropocene*, ed. Gregg Mittman, Marco Armiero, and Robert Emmett (Chicago: University of Chicago Press, 2018), 1–18; Anna Lowenhaupt Tsing, Andrew S. Mathews, and Nils Bubandt, "Patchy Anthropocene: Landscape Structure, Multispecies History, and the Retooling of Anthropology," *Current Anthropology* 60, no. 20 (2019): S186–S197; Kathyrn Yusoff, *A Billion Black Anthropocenes or None* (Minneapolis: University of Minnesota Press, 2018); Dipesh Chakrabarty, "Anthropocene Time," *History and Theory* 57, no. 1 (2018): 5–32.
34. Woodruff D. Smith, "Friedrich Ratzel and the Origins of Lebensraum," *German Studies Review* 3, no. 1 (1980): 53.
35. Mike Hulme, "Reducing the Future to Climate: A Story of Climate Determinism and Reductionism," *OSIRIS* 26 (2011): 246.
36. Yusoff, *A Billion Black Anthropocenes*, 12.

MICROCLIMATE ONE

1. Peter Sloterdijk, *Terror from the Air*, trans. Amy Patton and Steve Corcoran (Cambridge, MA: MIT Press, 2009), 108.

CHAPTER TWO. **INCLEMENT WEATHER**

1. Thomas K. Nakayama and Robert L. Krizek, "Whiteness: A Strategic Rhetoric," *Quarterly Journal of Speech* 81 (1995): 293; James Baldwin, "On Being White and Other Lies," in *The Cross of Redemption: Uncollected Writings*, ed. Randall Kenan (New York: Pantheon, 2010), 177.
2. Baldwin, "On Being White," 137.
3. Michael Omi and Howard Winant, *Racial Formations in the United States: From the 1960s to the 1990s*, 2nd ed. (New York: Routledge, 1994).
4. Sylvia Wynter, "Unsettling the Coloniality of Being/Power/Truth/Freedom: Towards the Human, after Man, Its Overrepresentation—an Argument," *New Centennial Review* 3, no. 3 (2003): 257–337; Denise Farreira da Silva, *Toward a Global Idea of Race* (Minneapolis: University of Minnesota Press, 2007).
5. Sexton is critiquing Mbembe's genealogy of necropolitics. Jared Sexton, "People-of-Color-Blindness: Notes on the Afterlife of Slavery," *Social Text* 28, no. 2 (103), 2010: 48. For works on the relation of Blackness to White coloniality, history, philosophy, ecology, and culture, see Achille Mbembe, *Critique of Black Reason*, trans. Laurent Dubois (Durham, NC: Duke University Press, 2017); Aimé Césaire, *Discourse on Colonialism*, trans. Joan Pinkham (New York: Monthly Review Press, 2000); Silva, *Toward a Global Idea of Race*; Denise Farreira da Silva, "Toward a Black Feminist Poethics: The Quest(ion) of Blackness toward the End of the World," *Black Scholar* 44, no. 2 (2014): 81–97; Frantz Fanon, *Black Skin, White Masks*, trans. Charles Lam Markmann (London: Pluto Press, 1986); Lewis R. Gordon, "Race, Theodicy, and the Normative Emancipatory Challenges of Blackness," *South Atlantic Quarterly* 112, no. 4 (2013): 725–36; Hortense J. Spillers, *Black, White and in Color: Essays on American Literature and Culture* (Chicago: University of Chicago Press, 2003); Kathryn Yusoff, *A Billion Black Anthropocenes or None* (Minneapolis: University of Minnesota Press, 2018); Calvin L. Warren, *Ontological Terror: Blackness, Nihilism, and Emancipation* (Durham, NC: Duke University Press, 2018); Frank B. Wilderson III, *Afropessimism* (New York: W. W. Norton, 2020); Wynter, "Unsettling the Coloniality."

6. Christina Sharpe, *In the Wake: On Blackness and Being* (Durham, NC: Duke University Press, 2016). Also see Wynter, "Unsettling the Coloniality"; Tiffany Lethabo King, *The Black Shoals: Offshore Formations of Black and Native Studies* (Durham, NC: Duke University Press, 2019); Wilderson, *Afropessimism*, 2020. Also see Eric King Watts, "Critical Cosmopolitanism, Antagonism, and Social Suffering," *Quarterly Journal of Speech* 101, no. 1 (2015): 271–79.
7. Watts, "Critical Cosmopolitanism," 277.
8. King, *The Black Shoals*, 18; Anne Anlin Cheng, *Ornamentalism* (New York: Oxford University Press, 2019); Kimberlé Crenshaw, "Mapping the Margins: Intersectionality, Identity Politics, and Violence against Women," in *Critical Race Theory: The Key Writings That Formed the Movement*, ed. Kimberlé Crenshaw, Neil Gotanda, Gary Pellar, and Kendall Thomas (New York: New Press, 1996), 357–83. See Wynter, "Unsettling the Coloniality," 2003.
9. Amy Brandzel, *Against Citizenship: The Violence of the Normative* (Urbana: University of Illinois Press, 2016), 60. Also see Frank B. Wilderson III, *Red, White & Black: Cinema and the Structure of U.S. Antagonisms* (Durham, NC: Duke University Press, 2010), 35–53.
10. For example, see Molefi Kete Asante, *Afrocentricity: The Theory of Social Change* (Chicago: African American Images, 2003); Amber E. Kelsie, "Blackened Debate at the End of the World," *Philosophy and Rhetoric* 52, no. 1 (2019): 63–70; Tammie M. Kennedy, Joyce Irene Middleton, and Krista Ratcliffe, *Rhetorics of Whiteness: Postracial Hauntings in Popular Culture, Social Media, and Education* (Carbondale: Southern Illinois University Press, 2017); Mark Lawrence McPhail, "A Question of Character: Re(-)signing the Racial Contract," *Rhetoric & Public Affairs* 7, no. 3 (2004): 391–405; Armond Towns, "Black 'Matter' Lives," *Women's Studies in Communication* 41, no. 4 (2018): 349–58; Watts, "Critical Cosmopolitanism"; Eric King Watts, "Postracial Fantasies, Blackness, and Zombies," *Communication and Critical/Cultural Studies* 14, no. 4 (2017): 317–33; Megan Eatman, *Ecologies of Harm* (Columbus: Ohio State University Press, 2020).
11. Nakayama and Krizek, "Whiteness," 295, 291.
12. Thomas K. Nakayama, "Whiteness Is Not Contained," *Communication and Critical/ Cultural Studies* 17, no. 2 (2020): 200, emphasis added.
13. Sharpe, *In the Wake*; Nakayama and Krizek, "Whiteness." A Google Ngram search indicates the first use of "racial climate" in a book was in 1829, with occasional references until the 1940s, and a steady increase in usage from the 1960s to today.
14. Darrell Wanzer-Serrano, "Rhetoric's Rac(e/ist) Problems," *Quarterly Journal of Speech*

105, no. 4 (2019): 465–76. Warren argues "that black ~~being~~ constitutes the nothing in an antiblack world, which is continually degraded, dominated, and violated. Antiblackness is anti-nothing. . . . By abandoning the human, human-ness, and the liberal humanism that enshrouds it, we can better understand the violent formations of antiblackness, particularly ontological terror." *Ontological Terror*, 169–70. For examples of discussion of Black fugitivity in the face of ontological terror, see Sharpe, *In the Wake*; Nahum Dmitri Chandler, *X—The Problem of the Negro as a Problem for Thought* (New York: Fordham University Press, 2014); Silva, "Toward a Black Feminist Poethics"; Gordon, "Race, Theodicy"; Fred Moten, *In the Break: The Aesthetics of the Black Radical Tradition* (Minneapolis: University of Minnesota Press, 2003); Fred Moten, *Black and Blur* (Durham, NC: Duke University Press, 2017); Fred Moten, *Stolen Life* (Durham, NC: Duke University Press, 2018); Saidiya V. Hartman, *Wayward Lives, Beautiful Experiments: Intimate Histories of Social Upheaval* (New York: W.W. Norton, 2019). Wilson has recently argued that "furtiveness" accounts for Black fugitive life in relation to law as "already illegal and lashable." See T. Anansi Wilson, "Furtive Blackness: On Blackness and Being," *Hastings Constitutional Law Quarterly* 48, no. 1 (2020): 144.

15. Édouard Glissant, *The Poetics of Relation*, trans. Betsy Wing (Ann Arbor: University of Michigan Press, 2010); Alexander G. Weheliye, *Habeas Viscus: Racializing Assemblages, Biopolitics, and Black Feminist Theories of the Human* (Durham, NC: Duke University Press, 2014); Mbembe, *Critique of Black Reason*.
16. Joseph Pugilese, "Biometrics, Infrastructural Whiteness, and the Racialized Zero Degree of Nonrepresentation," *boundary 2* 34, no. 2 (2007): 107, 110.
17. W. E. B. Du Bois, *The Souls of Black Folk* (G&D Media, 2019); King, *The Black Shoals*; Césaire, *Discourse on Colonialism*; Fanon, *Black Skin, White Masks*; Warren, *Ontological Terror*; Hortense Spillers, "Mama's Baby, Papa's Maybe: An American Grammar Book," in *The Black Feminist Reader*, ed. Joy James and T. Denean Sharpley-Whiting (Malden, MA: Blackwell, 2000), 57–87; Yusoff, *A Billion Black Anthropocenes*; Leilani Nishime and Kim D. Hester Williams, *Racial Ecologies* (Seattle: University of Washington Press, 2018); Gordon, "Race, Theodicy"; Arjun Appadurai, *Fear of Small Numbers: An Essay on the Geography of Anger* (Durham, NC: Duke University Press, 2006); Silva, *Toward a Global Idea of Race*; Moten, *Stolen Life*; Fred Moten, *Universal Machine* (Durham, NC: Duke University Press, 2018).
18. Katherine McKittrick, *Demonic Grounds: Black Women and the Cartographies of Struggle* (Minneapolis: University of Minnesota Press, 2006); Sharpe, *In the Wake*, 108–13; also see Ersula J. Ore, *Lynching: Violence, Rhetoric, and American Identity*

(Jackson: University Press of Mississippi, 2019), 101–21, 133–36; Fanon, *White Skin, Black Masks*; Spillers, "Mama's Baby, Papa's Maybe"; King, *The Black Shoals*.

19. Silva, *Toward a Global Idea of Race*.
20. Rachel Gorman, "Disablement In and For Itself: Toward a 'Global' Idea of Disability," *Somatechnics* 6, no. 2 (2016): 257.
21. Saidiya V. Hartman, *Scenes of Subjection: Terror, Slavery, and Self-Making in Nineteenth-Century America* (New York: Oxford University Press, 1997), 21, 25–26. In *The Black Shoals*, King notably elaborates on Black fungibility relative to settler colonialism and Indigenous life.
22. Zakiyyah Iman Jackson, *Becoming Human: Matter and Meaning in an Antiblack World* (New York: NYU Press, 2020), 35.
23. Jackson, *Becoming Human*, 198. Also see King, *The Black Shoals*, 49.
24. Yanar Hashlamon, "Rhetoricity at the End of the World: Defining Rhetorical Debility under Neoliberal Capitalism," *Rhetoric Society Quarterly* 52, no. 1 (2022): 23; Jasbir K. Puar, *The Right to Maim: Debility, Capacity, Disability* (Durham, NC: Duke University Press, 2017).
25. McKittrick, *Demonic Grounds*, x.
26. McKittrick, *Demonic Grounds*, 7.
27. Katherine McKittrick, "Plantation Futures," *Small Axe* 42 (November 2013): 7.
28. Thomas Rickert, *Ambient Rhetoric: The Attunements of Rhetorical Being* (Pittsburgh: University of Pittsburgh Press, 2013); Nakayama and Krizek, "Whiteness," 291.
29. Silva, *Toward a Global Idea of Race*, 71–76.
30. Jackson, *Becoming Human*, 208. Also see King, *The Black Shoals*, 45.
31. Rickert, *Ambient Rhetoric*, 258.
32. Watts, "Critical Cosmopolitanism," 277. Watts asks us to "enjoin an analysis of the *vocabulary* of impossibility itself" (emphasis added). Following Sharpe, I am thinking beyond vocabulary to order and action more generally.
33. Sharpe, *In the Wake*, 104.
34. Sharpe, *In the Wake*, 111.
35. Nakayama and Krizek, "Whiteness," 291.
36. Sharpe, *In the Wake*, 106.
37. Sharpe, *In the Wake*, 13–14.
38. Sharpe, *In the Wake*, 21–22, 13.
39. Sharpe, *In the Wake*, 19, 21; also see Judith Butler, *The Force of Nonviolence: An Ethico-Political Bind* (London: Verso, 2020).
40. A climate is difficult to atone for (in contrast to a contract) because it defies the

resources of moral reasoning or discourse that are suited to reconciliation or reparation. McPhail, "A Question of Character," 401.

41. NASA, "NASA—What's the Difference Between Weather and Climate?," NASA, February 1, 2005, https://www.nasa.gov.
42. Lisa A. Flores, *Deportable and Disposable: Public Rhetoric and the Making of the 'Illegal' Immigrant* (University Park: Penn State University Press, 2020), 9.
43. Flores, *Deportable and Disposable*, 7, 8.
44. Flores, *Deportable and Disposable*, 16.
45. Flores, *Deportable and Disposable*, 66.
46. Flores, *Deportable and Disposable*, 51.
47. Flores, *Deportable and Disposable*, 63.
48. Flores, *Deportable and Disposable*, 63, 57.
49. Sharpe, *In the Wake*, 116.
50. Sharpe, *In the Wake*, 20, 5.
51. Gorman, "Disablement," 257.
52. Nirmala Erevelles, *Disability and Difference in Global Contexts: Enabling a Transformative Body Politic* (New York: Palgrave Macmillan, 2011), 26. Also see pages 33–38. See Jay Timothy Dolmage, *Disability Rhetoric* (Syracuse, NY: Syracuse University Press, 2014), 22–29.
53. Sharpe, *In the Wake*, 7.
54. Sharpe, *In the Wake*, 4, 21.
55. Sharpe, *In the Wake*, 107, 15.
56. Nathan Stormer, "Afterword: Working in an Ecotone," in *Tracing Rhetoric and Material Life: Ecological Approaches*, ed. Bridie McGreavy, Justine Wells, George F. McHendry Jr., and Samantha Senda-Cook (New York: Palgrave Macmillan, 2018), 343–54.
57. Diane M. Keeling and Jennifer C. Prairie, "Trophic and Tropic Dynamics: An Ecological Perspective on Tropes," in *Tracing Rhetoric and Material Life: Ecological Approaches*, ed. Bridie McGreavy, Justine Wells, George F. McHendry Jr., and Samantha Senda-Cook (New York: Palgrave Macmillan, 2018), 50–51.
58. Caroline Gottschalk Druschke, "A Trophic Future for Rhetorical Ecologies," *Enculturation* (February 10, 2019).
59. Puar, *The Right to Maim*, 10–18. In that sense, trophism provides an additional, alternate frame for discussing the consumption otherness. For example, bell hooks famously described the risk of media representations of difference in the face of the commodification of otherness for White consumption: "The over-riding fear is that

cultural, ethnic, and racial differences will be continually commodified and offered up as new dishes to enhance the white palate—that the Other will be eaten, consumed, forgotten." See bell hooks, *Black Looks: Race and Representation* (Boston, MA: South End, 1992), 39. Trophism would press us to consider consumption on material terms that go beyond White viewing and pleasure.

60. Druschke, "A Trophic Future."
61. Eatman, *Ecologies of Harm*, 138.
62. Eatman, *Ecologies of Harm*, 9.
63. Eatman, *Ecologies of Harm*, 13.
64. Eatman, *Ecologies of Harm*, 139.
65. Eatman, *Ecologies of Harm*, 149.
66. King, *The Black Shoals*, 10. The vampiric quality of Whiteness, namely as an appropriative and consumptive relation to otherness, is well understood. Jordan Peele's *Get Out* is a horror masterpiece of that very idea.
67. Sharpe, *In the Wake*, 16.
68. Sharpe, *In the Wake*, 106.
69. Sharpe, *In the Wake*, 30.
70. Sharpe, *In the Wake*, 124.
71. King, *The Black Shoals*, 40. King understands colonial conquest using a white-black-red triangle to analyze its relations, but similarly describes it as a milieu.
72. Armond Towns, "Rebels of the Underground: Media, Orality, and the Routes of Black Emancipation," *Communication and Critical/Cultural Studies* 13, no. 2 (2016): 184–97.
73. See Kara Keeling, *Queer Times, Black Futures* (New York: NYU Press, 2019); and Tavia Nyong'o, *Afro-Fabulations: The Queer Drama of Black Life* (New York: NYU Press, 2019); also see Rickert, *Ambient Rhetoric*, 239.
74. Sharpe, *In the Wake*, 106. Also see Silva, "Toward a Black Feminist Poethics," on Black feminist ethical poesis, or "poethics"; and Moten, *Stolen Life*, 2018.
75. Eatman, *Ecologies of Harm*, 138.
76. I adapt Rickert's example of the gasoline-powered car; the language is resonant although the case is not analogous (*Ambient Rhetoric*, 264).
77. Sharpe, *In the Wake*, 20.
78. Radha Blank recently wrote and directed a contemporary comedic commentary on this long-standing truth of White media ownership (*The Forty-Year-Old Version*, 2020). Also see Paul C. Taylor, *Black Is Beautiful: A Philosophy of Black Aesthetics* (Hoboken, NJ: Wiley, 2016), 49–58.

79. Olga Idriss Davis, "A Black Woman as Rhetorical Critic: Validating Self and Violating the Space of Otherness," *Women's Studies in Communication* 21, no. 1 (1998): 85–86; Lisa A. Flores, "Between Abundance and Marginalization: The Imperative of Racial Rhetorical Criticism," *Review of Communication* 6, no. 1 (2015): 16–17; Wanzer-Serrano, "Rhetoric's Rac(e/ist) Problems." Spillers described it thusly: "the academy offer the black creative intellectual his own, dear laboratory." Hortense J. Spillers, *Black, White, and in Color: Essays on American Literature and Culture* (Chicago: University of Chicago Press, 2003), 469. Also see Chandler, *X—The Problem of the Negro.*
80. Wizard Marks, "I–35W Disrupted Minority Community, Boxed-In Phillips," *The Alley*, August 1990; National Resource Defense Council, "Flint Water Crisis: Everything You Need to Know," National Resource Defense Council, November 8, 2018, https://www.nrdc.org/stories/flint-water-crisis-everything-you-need-know.
81. Harriet A. Washington, *Medical Apartheid: The Dark History of Medical Experimentation on Black Americans from Colonial Times to the Present* (New York: Harlem Moon, 2008).
82. Watts, "Postracial Fantasies," 317–33.
83. Taylor, *Black Is Beautiful.*
84. Wilderson, *Afropessimism*, 188–228.
85. This is part of what Spillers describes as "pornotroping" ("Mama's Baby," 60). Also see Weheliye, *Habeas Viscus*, 89–112.
86. Eve Tuck and K. Wayne Yang, "Decolonization Is Not a Metaphor," *Decolonization: Indigeneity, Education & Society* 1, no. 1 (2012): 1–40.
87. Sharpe, *In the Wake*, 17.
88. Druschke, "A Trophic Future."
89. Rickert, *Ambient Rhetoric*, 239. This differs from but complements critiques of rhetoric's embodiment, such as Dolmage's (*Disability Rhetoric*, 2014) analysis of the field of rhetoric's disavowed, normate body. Climatic thinking as I am discussing it is not about embodiment as a rhetorical project *per se* but it does open possibilities for understanding the history of the human's embodiment as a *matrix* for global relationality; that requires unbinding rhetoric from the human in the last instance, however.
90. Yusoff, *A Billion Black Anthropocenes*, 104.

CHAPTER THREE. **CLIMATES OF BENEVOLENT ABLEISM**

1. Stella Young, "I'm Not Your Inspiration, Thank You Very Much," TEDxSydney, April 2014, video, 9.03, posted June 9, 2014, https://www.ted.com/talks/.
2. Katie R. G. Pryal, "The Genre of the Mood Memoir and the Ethos of Psychiatric Disability," *Rhetoric Society Quarterly* 40, no. 5 (2010): 481.
3. Jan Grue, "The Problem with Inspiration Porn: A Tentative Definition and a Provisional Critique," *Disability & Society* 31, no. 6 (2016): 847.
4. Katie Ellis, *Disability and Popular Culture: Focusing Passion, Creating Community and Expressing Defiance* (New York: Routledge, 2016), 148, 158.
5. Young, "I'm Not Your Inspiration," 2014.
6. James L. Cherney, *Ableist Rhetoric: How We Know, Value, and See Disability* (State College: Penn State University Press, 2019), 8.
7. Simi Linton, *Claiming Disability: Knowledge and Identity* (New York: NYU Press, 1998), 11.
8. Aimée Morrison, "(Un)reasonable, (Un)necessary, and (In)appropriate: Biographic Mediation of Neurodivergence in Academic Accommodations," *Biography* 42, no. 2 (2019): 693–719.
9. Imani Barbarin (@Imani_Barbarin), "I think about the time an abled random stranger threw my crutch into the pool 'to help me swim' a lot," Twitter, March 15, 2019, 6:14 p.m.
10. Ellen W. Gorsevski, *Peaceful Persuasion: The Geopolitics of Nonviolent Rhetoric* (Albany: State University of New York Press, 2004), 128.
11. Christina Sharpe, *In the Wake: On Blackness and Being* (Durham, NC: Duke University Press, 2016), 104.
12. Saidiya V. Hartman, *Wayward Lives, Beautiful Experiments: Intimate Histories of Social Upheaval* (New York: W.W. Norton, 2019), xiv.
13. Sharpe, *In the Wake*.
14. Carolyn R. Miller, Amy J. Devitt, and Victoria J. Gallagher, "Genre: Permanence and Change," *Rhetoric Society Quarterly* 38, no. 3 (2018): 273.
15. Christina V. Cedillo, "What Does It Mean to Move? Race, Disability, and Critical Embodiment Pedagogy," *Composition Forum* 39 (2018).
16. Jay Timothy Dolmage, *Disabled upon Arrival: Eugenics, Immigration, and the Construction of Race and Disability* (Columbus: Ohio State University Press, 2018), 31.
17. Cherney, *Ableist Rhetoric*, 48–49.

18. Mel Y. Chen, *Animacies: Biopolitics, Racial Mattering, and Queer Affect* (Durham, NC: Duke University Press, 2012), 181, 184.
19. Sara J Newman, "Disability and Life Writing: Reports from the Nineteenth-Century Asylum," *Journal of Literary & Cultural Disability Studies* 5, no. 3 (2011): 265.
20. Rosemarie Garland-Thomson, *Extraordinary Bodies: Figuring Physical Disability in American Culture and Literature* (New York: Columbia University Press, 1997), 63.
21. Even with this increased awareness, there are too many examples to list of using physical disability and/or difference as a choreographed sign of a character's ill intent. A recent one from television is that of Barry Kripke on CBS sitcom *The Big Bang Theory*; as the foil to the show's protagonists, the comic contours of this character as an antagonist were crudely flagged through the actor's performance of an exaggerated lisp.
22. Sara Ahmed critiques how representation in the name of diversity can quickly become absorbed into the background of institutions. See Sara Ahmed, *On Being Included: Racism and Diversity in Institutional Life* (Durham, NC: Duke University Press, 2012).
23. Michelle Friedner and Karen Weingarten, "Introduction: Disorienting Disability," *South Atlantic Quarterly* 118, no. 3 (2019): 484.
24. Garland-Thomson, *Extraordinary Bodies*, 7.
25. Ria Cheyne, *Disability, Literature, Genre: Representation and Affect in Contemporary Fiction* (Liverpool: Liverpool University Press, 2019), 4.
26. Rick Carpenter, "Disability as Socio-Rhetorical Action: Towards a Genre-Based Approach," *Disability Studies Quarterly* 31, no. 3 (2011).
27. Sami Schalk, "Reevaluating the Supercrip," *Journal of Literary & Cultural Disability Studies* 10, no. 1 (2016): 74.
28. Sami Schalk, "Black Disability Gone Viral: A Critical Race Approach to Inspiration Porn," *College Language Association Journal* 64, no. 1 (2021): 101.
29. Schalk, "Black Disability," 101.
30. Eli Clare, *Exile and Pride: Disability, Queerness, and Liberation* (Durham, NC: Duke University Press, 2015), 2.
31. Garland-Thomson, *Extraordinary Bodies*, 65. Emphases added.
32. Jordan Culver, "4th grader with Spina Bifida Almost Couldn't Attend a Class Trip," *USA Today*, September 24, 2019. "Teen Couple with Down's Syndrome Voted Prom King and Queen," *BBC News*, July 16, 2019.
33. Rosemarie Garland-Thomson, "Seeing the Disabled: Visual Rhetorics of Disability in Popular Photography," in *New Disability History*, ed. Paul K. Longmore and Lauri

Umansky (New York: NYU Press, 2001), 339, 340.

34. Grue, "The Problem with Inspiration Porn," 841–42.
35. Sara Ahmed, *The Cultural Politics of Emotion*, 2nd ed. (Edinburgh: Edinburgh University Press, 2014), 45.
36. See Ryan Skinnell, "Using Democracy against Itself: Demagogic Rhetoric as an Attack on Democratic Institutions," *Rhetoric Society Quarterly* 49, no. 3 (2019): 248–63.
37. Allyson Chiu, "A Fox News Guest Called Greta Thunberg 'Mentally Ill': The Network Apologized for the 'Disgraceful' Comment," *Washington Post*, September 24, 2019.
38. Tanya Lewis, "The 'Shared Psychosis' of Donald Trump and His Loyalists," *Scientific American*, January 11, 2021.
39. Sami Schalk (@DrSamiSchalk), "This discursive approach is ableist bc it suggests that people with psych disabilities are inherently dangerous, dismissible, disposable, etc. It adds to the oppression of folks with psych disabilities, relies on this oppression to be effective," Twitter, April 25, 2020, 12:33 p.m.
40. Schalk, "Reevaluating," 74.
41. Tiffany Lethabo King, *The Black Shoals: Offshore Formations of Black and Native Studies* (Durham, NC: Duke University Press, 2019), 20. Rashawn Ray, "Why Are Blacks Dying at Higher Rates from COVID-19?," *Brookings*, April 9, 2020.
42. Kevin Roose, "Sorry, but Working from Home Is Overrated," *New York Times*, March 10, 2020.
43. Alabama Department of Public Health, "Annex to ESF 8 of the State of Alabama Emergency Operations Plan," 2010, Alabama Department of Public Health, https://www.alabamapublichealth.gov/, 8. After review from the Office of Civil Rights, the Alabama Department of Public Health agreed to take the plan down from all websites and publicly disavow this policy in April 2020.
44. M. Remi Yergeau, *Authoring Autism: On Rhetoric and Neurological Queerness* (Durham, NC: Duke University Press, 2018), 56.
45. Nathan Stormer and Bridie McGreavy, "Thinking Ecologically about Rhetoric's Ontology: Capacity, Vulnerability, and Resilience," *Philosophy and Rhetoric* 50, no. 1 (2017): 2.
46. Jay Timothy Dolmage, "Metis, *Mêtis*, *Mestiza*, Medusa: Rhetorical Bodies across Rhetorical Traditions," *Rhetoric Review* 28, no. 1 (2009): 21.
47. Petra Kuppers, "Toward a Rhizomatic Model of Disability: Poetry, Performance, and Touch," *Journal of Literary and Cultural Disability Studies* 3, no. 3 (2009): 225–26.

48. Amy Vidali, "Texts of Our Institutional Lives: Performing the Rhetorical Freak Show—Disability, Student Writing, and College Admissions," *College English* 69, no. 6 (2007): 618, 619.

MICROCLIMATE THREE

1. Max Liboiron, *Pollution Is Colonialism* (Durham, NC: Duke University Press, 2021), 155.
2. Naomi Ortiz, "Majestic Disabled/Queer/People of Color Elders Instruct How to Dance in the Struggle," EcoSemantics Archive: Field Notes from the 2020 EcoSomatics Symposium, *Jacket2*, May 6, 2022.

CHAPTER FOUR. **DISRUPTING ENVIRONMENTAL IMAGINATION, TOWARD A TIDAL ETHICS**

1. Sheila Jasanoff, "Image and Imagination: The Formation of Global Environmental Consciousness," in *Science and Public Reason* (London: Taylor & Francis Group, 2012), 78, 79. Jasanoff's essay was originally published in *Changing the Atmosphere: Expert Knowledge and Environmental Governance*, ed. Paul Edwards and Clark Miller (Cambridge, MA: MIT Press, 2001), 309–37. Emphasis added.
2. Jasanoff, "Image and Imagination," 98.
3. Jasanoff, "Image and Imagination," 85.
4. John Beardsley, "The Massachusetts Body of Liberties," General Court of the Commonwealth of Massachusetts Bay, December, 1641, Winthrop Society, updated 2015, https://www.winthropsociety.com/.
5. Édouard Glissant, *The Poetics of Relation*, trans. Betsy Wing (Ann Arbor: University of Michigan Press, 2010), 144–51.
6. Tiffany Lethabo King, *The Black Shoals: Offshore Formations of Black and Native Studies* (Durham, NC: Duke University Press, 2019), 29.
7. Glissant, *Poetics*, 199.
8. Robert Poole, *Earthrise: How Man First Saw the Earth* (New Haven, CT: Yale University Press, 2010), 142.
9. Ian Sample, "Earthrise: How the Iconic Image Changed the World," *Guardian*, December 24, 2018.

10. Jasanoff, "Image and Imagination, 85.
11. Jasanoff, "Image and Imagination," 85.
12. Jasanoff, "Image and Imagination," 91.
13. Poole, *Earthrise*, 11.
14. Jasanoff, "Image and Imagination," 91.
15. Poole, *Earthrise*. This quote draws from the book of Genesis reading that the Apollo 8 astronauts performed as they witnessed the lunar sunrise on Christmas Eve. The passage starts to point to how earth images from space were and continue to be articulated in distinctly racialized ways. As Zakiyyah Iman Jackson argues, the linkages between darkness and void are expressions of the Black mater as a gendered matrix of anti-Blackness. See Zakiyyah Iman Jackson, *Becoming Human: Matter and Meaning in an Antiblack World* (New York: NYU Press, 2020).
16. Kevin Michael DeLuca and Anne Teresa Demo, "Imaging Nature: Watkins, Yosemite, and the Birth of Environmentalism," *Critical Studies in Media Communication* 17, no. 3 (2000): 242.
17. Deluca and Demo, "Imaging Nature," 251.
18. Deluca and Demo, "Imaging nature," 256.
19. Max Liboiron, *Pollution Is Colonialism* (Durham, NC: Duke University Press, 2021), 11.
20. Nathan Stormer, "Articulation: A Working Paper on Rhetoric and Taxis," *Quarterly Journal of Speech* 90, no. 3 (2004): 274.
21. Stormer, "Articulation," 275.
22. Laurie Gries, *Still Life with Rhetoric: A New Materialist Approach for Visual Rhetorics* (Logan: Utah State University Press, 2015), 28.
23. Kevin Michael DeLuca, *Image Politics: The New Rhetoric of Environmental Activism* (New York: Routledge, 2005), 37–38.
24. Stormer, "Articulation," 271.
25. Kim TallBear, *Native American DNA: Tribal Belonging and the False Promise of Genetic Science* (Minneapolis: University of Minnesota Press, 2013).
26. Jackson, *Becoming*, 122.
27. Jackson, *Becoming*, 69.
28. Kathryn Yusoff, *A Billion Black Anthropocenes or None* (Minneapolis: University of Minnesota Press), 3–11.
29. TallBear, *Native American DNA*, 23–29, 176.
30. Kyle Powys Whyte, "Against Crisis Epistemology," in *Handbook of Critical Indigenous Studies*, ed. Brendan Hokowhitu, Aileen Moreton-Robinson, Linda Tuhiwai-Smith,

Chris Andersen, and Steve Larkin (New York: Routledge, 2021), 53–57.

31. Jasanoff, "Image and Imagination," 86.
32. Julie Doyle, *Mediating Climate Change* (New York: Routledge, 2011), 22. Whyte, "Against Crisis Epistemology," 53.
33. Whyte, "Against Crisis Epistemology," 54.
34. Whyte, "Against Crisis Epistemology," 53, 55–57.
35. Leah Ceccarelli, *On the Frontier of Science: An American Rhetoric of Exploration and Exploitation* (East Lansing: Michigan State University Press, 2013), 31–49.
36. Poole, *Earthrise*, references to Columbus appear on pages 6, 7, 20, 34, 42, 56, 61, and 127.
37. Poole, *Earthrise*, 6.
38. Poole, *Earthrise*, 34.
39. Yusoff, *A Billion Black Anthropocenes*, 14, 33–39.
40. Bridie McGreavy, Darren Ranco, John Daigle, Suzanne Greenlaw, Nolan Altvater, Tyler Quiring, Natalie Michelle, Jan Paul, Maliyan Binette, Brawley Benson, Anthony Sutton, and David Hart, "Science in Indigenous Homelands: Addressing Power and Justice in Sustainability Science from/with/in the Penobscot River," *Sustainability Science* 16, no. 3 (2021): 942.
41. Barbara Adam, *Timescapes of Modernity: The Environment and Invisible Hazards* (London: Routledge, 2005).
42. Poole, *Earthrise*, 1.
43. Poole, *Earthrise*, 72.
44. Bill Steigerwald, "NASA Releases New High-Resolution Earthrise Image," NASA, updated August 7, 2017, https://www.nasa.gov/image-feature/goddard/lro-earthrise-2015.
45. Tema Milstein, "The Performer Metaphor: 'Mother Nature Never Gives Us the Same Show Twice,'" *Environmental Communication* 10, no. 2 (2016): 238. Milstein draws from Phil Macnaghten and John Urry, *Contested Natures* (London: SAGE, 1998).
46. Jackson, *Becoming*, 27.
47. Jackson, *Becoming*, 42.
48. Jackson, *Becoming*, 39.
49. Jackson, *Becoming*, 100; Glissant, *Poetics*, 26.
50. Jackson, *Becoming*, 166.
51. Yusoff, *A Billion Black Anthropocenes*, 26–28.
52. David R. Williams, "The Apollo 8 Christmas Eve Broadcast," NASA Goddard

Space Flight Center, updated September 25, 2007, https://nssdc.gsfc.nasa.gov/planetary/lunar/apollo8_xmas.html.

53. Archibald MacLeish, "Riders on Earth Together, Brothers in Eternal Cold," *New York Times*, December 25, 1968, 1.
54. Jackson, *Becoming*, 83.
55. Glissant, *Poetics*, 51.
56. Glissant, *Poetics*, 48.
57. Jackson, *Becoming*, 86.
58. Glissant, *Poetics*, 196.
59. Glissant, *Poetics*, 196.
60. See Bridie McGreavy, Sara Randall, Tyler Quiring, Carter Hathaway, and Gabrielle Hillyer, "Enhancing Adaptive Capacities in Coastal Communities through Engaged Communication Research: Insights from a Statewide Study of Shellfish Co-Management," *Ocean & Coastal Management* 163 (2018): 240–53; and McGreavy et al., "Science in Indigenous Homelands."
61. Glissant, *Poetics*, 190.
62. I want to thank Casey Boyle for reviewing an earlier version of this essay and providing the suggestion to add this photo as a way to set up a contrast with earth images and create a feel for tidal places.
63. In discussing reciprocity, TallBear draws a distinction between giving back and "standing with," the latter of which intends a feminist orientation to an ethics of accountability. In this approach, listening becomes the means through which knowledge coproduction occurs and where the research focuses on addressing systems of power as opposed to focusing on "research subjects" themselves. See Kim TallBear, "Standing with and Speaking as Faith: A Feminist-Indigenous Approach to Inquiry," *Journal of Research Practice* 10, no. 2 (2014): 4.
64. Liboiron, *Pollution Is Colonialism*, 133.
65. See Tiara R. Na'puti, "Archipelagic Rhetoric: Remapping the Marianas and Challenging Militarization from 'A Stirring Place,'" *Communication and Critical/Cultural Studies* 16, no. 1 (2019): 4–25. Na'puti defines colonial cartographic violence as "the textual, linguistic, visual, and material dynamics constructing and registering places as existing exclusively for colonization and militarization. Colonial naming and mapping processes privilege landmasses over expansive seas, islands, and archipelagos," 3.
66. King, *The Black Shoals*, 27, citing her dissertation, "In the Clearing: Black Female Bodies, Space and Settler Colonial Landscapes" (PhD diss., University of Maryland,

College Park, 2013), 13.

67. King, *The Black Shoals*, 28.
68. King, *The Black Shoals*, 28.
69. Eve Tuck and K. Wayne Yang, "Decolonization Is Not a Metaphor," *Decolonization: Indigeneity, Education & Society* 1, no. 1 (2012): 35–36.

MICROCLIMATE FOUR

1. Anna Lowenhaupt Tsing, *The Mushroom at the End of the World* (Princeton: Princeton University Press, 2015), 28.

CHAPTER FIVE. INSTITUTIONAL CLIMATE CHANGING

1. The definitions that start each section in my essay are original, so to speak, but also collective in the sense that they were inspired by the books we read and our group's conversations about climate, as well as from my investigation of the genealogy of climate terms.
2. The place I work, the University of Washington–Seattle, is placed on the unceded ancestral lands of the Coast Salish peoples, the land which touches the shared waters of all tribes and bands within the Duwamish, Puyallup, Suquamish, Tulalip, and Muckleshoot nations. I am a grateful but uninvited guest on these lands.
3. Some questions used in these literacy tests would be laughable if not for their malevolence. The full egregiousness of the questions first hit when I read a poem by Joel Felix, who was working with found language that included one designed to disenfranchise Black voters: *How many bubbles in a bar of soap?* I mean, how's that for the so-called neutrality of literacy exams? See Catherine Prendergast, *Literacy and Racial Justice: The Politics of Learning after Brown v. Board of Education* (Carbondale: Southern Illinois University Press, 2003); and Scott Richard Lyons, "Rhetorical Sovereignty: What Do American Indians Want from Writing?," *College Composition and Communication* 51, no. 3 (2000): 447–68.

   See April Baker-Bell, "*We Been Knowin*: Toward an Antiracist Language & Literacy Education," *Journal of Language and Literacy Education* 16, no. 1 (Spring 2020); Asao Inoue, *Antiracist Writing Assessment Ecologies: Teaching and Assessing Writing for a Socially Just Future* (Fort Collins, CO: WAC Clearinghouse, 2015);

Rosina Lippi-Green, *English with an Accent: Language, Ideology and Discrimination in the United States* (New York: Routledge, 2012); Suhanthie Motha, *Race, Empire, and English Language Teaching: Creating Responsible and Ethical Anti-Racist Practice* (New York: Teachers College Press, 2014); la paperson, *A Third University Is Possible* (Minneapolis: University of Minnesota, 2017); and many others.

4. Prendergast, *Literacy and Racial Justice*, 2, 7.
5. Motha, *Race, Empire, and English Language Teaching*, 37.
6. Eduardo Bonilla-Silva, *Racism without Racist: Color-Blind Racism and the Persistence of Racial Inequality in American* (Lanham, MD: Rowman & Littlefield, 2013), quoted in Motha, *Race, Empire, and English Language Teaching*, 37–38.
7. Inoue, *Antiracist Writing Assessment Ecologies, 8.*
8. Inoue, *Antiracist Writing Assessment Ecologies, 8.*
9. See April Baker-Bell, *Linguistic Justice: Black, Language, Literacy, Identity, and Pedagogy* (Urbana, IL: National Council of Teachers of English, 2020); Juan C. Guerra, "An Embodied History of Language Ideologies," in *Writing Across Difference: Theory and Intervention*, ed. James R. Daniel, Katie Malcolm, and Candice Rai (Logan: Utah State University Press, 2022), 1–38; Inoue, *Antiracist Writing Assessment Ecologies*; Django Paris and H. Samy Alim, eds., *Culturally Sustaining Pedagogy: Teaching and Learning for Justice in a Changing World* (New York: Teachers College Press, 2017); and Sumyat Thu, Katie Malcolm, Candice Rai, and Anis Bawarshi, "Antiracist Translingual Praxis in Writing Ecologies," in *Writing Across Difference: Theory and Intervention*, ed. James Rushing Daniel, Katie Malcolm, and Candice Rai (Logan: Utah University Press, 2022), 195–217.
10. PWR supports about seventy-five instructors who teach roughly 240 sections of composition to 5,500 students annually.
11. Unpacking this "we" is not simple. Climates have vast, complex relationships among so many people, things, ideas, and all manner else. Regarding the work described here, I especially acknowledge Sumyat Thu, Bell Kim, Denise Grollmus, Holly Shelton, Nolie Ramsey, Lydia Heberling, Emily George, Matthew Hitchman, Anselma Prihandita, A. J. Burgin, Nanya Jhingran, Liz Jansen, Brittney Frantece, Sara Lovett, T. J. Walker, MiSun Bishop, Alycia Gilbert, Jacki Fiscus, Ann Shivers-McNair, Alex Meany, Olivia Hernandez, Joe Concannon, Alec Fisher, Sarah Moore, Alex Smith, Yasmine Romero, Anis Bawarshi, Juan Guerra, Nancy Bou Ayash, Suhanthie Motha, Elizabeth Simmons-O'Neill, Michelle Liu, Kimberlee Gillis-Bridges, and Jake Huebsch. Beyond this, there are many other graduate student assistant directors, more than three hundred instructors, some forty thousand

students, colleagues, and many campus partners we worked with in this time. When I say this work was collective, I cannot overemphasize that point. This was "we" work, not me work.

12. Staci M. Perryman-Clark and Collin Craig, "Black Matters: Writing Program Administration in Twenty-First-Century Higher Education," in *Black Perspectives in Writing Program Administration: From the Margins to the Center*, ed. Staci M Perryman-Clark and Collin Craig (Urbana, IL: National Council of Teachers of English, 2019), 9, 11.
13. Sara Ahmed, *On Being Included: Racism and Diversity in Institutional Life* (Durham, NC: Duke University Press, 2012), 156.
14. Sharpe, *In the Wake: On Blackness and Being* (Durham, NC: Duke University Press, 2016), 104.
15. Sharpe, *In the Wake*, 18.
16. Sharpe, *In the Wake*, 130–31.
17. Baker-Bell, *Linguistic Justice*, 2–3.
18. Ahmed, *On Being Included*, 26.
19. Ahmed, *On Being Included*, 26.
20. Ahmed, *On Being Included*, 24.
21. Ahmed, *On Being Included*, 27.
22. Also known as K. Wayne Yang.
23. La paperson, *A Third University*, xiii.
24. La paperson, *A Third University*, xiv.
25. La paperson, *A Third University*, xiv.
26. On the matter of "kicking back," Kenneth Burke's genealogical thinking on *recalcitrance* has come to my mind over the years when thinking about institutional change. If the commonplace understanding of recalcitrance is something like stubbornness, immobility (a word implying something stanchly stuck-in-its-place), the genealogy of recalcitrance not only reveals a link to "heel" (e.g., foot part) but also indexes an active kicking back. Thinking of the recalcitrancy of institutions as actively "kicking back" is to see institutions not so much as inert or static but as complex climates energetically kicking themselves back into place every which way you turn. Fuck.
27. Liboiron, *Pollution Is Colonialism*, 21.
28. Raewyn Connell, *Southern Theory: The Global Dynamics of Knowledge in Social Science* (Boston: Polity, 2007), quoted in Liboiron, *Pollution Is Colonialism*, 21.
29. Michelle Murphy, "Against Population, toward Alterlife," in *Making Kin, Not*

Population, ed. Adele E. Clarke and Donna Jeanne Haraway (Chicago: Prickly Paradigm Press, 2018), 122–23.

30. Sara Ahmed, *Willful Subjects* (Durham, NC: Duke University Press, 2014), 133.
31. Ahmed, *Willful Subjects*, 143.
32. Ahmed, *Willful Subjects*, 143.
33. "PWR Statement on Antiracist Writing Pedagogy and Program Praxis," Program in Writing & Rhetoric, Department of English, University of Washington–Seattle, 2018, revised July 2021, https://english.washington.edu/diversity-equity-and-justice.
34. Sumyat Thu deserves the top credit here for her scholarly expertise, her personal lived and community-engaged experiences, and her fearless political and ethical convictions, all of which centrally guided and informed this work. I played a very active role in crafting and revising, too, but perhaps even more so in coordinating a large team to get cracking at the endless ocean of labor required to live by our commitments and make changes in our complex program. Other contributors include Anis Bawarshi, Alycia Gilbert, Emily George, Brittney Frantece, Stephanie Kerschbaum, Michelle Liu, Katie Malcolm, Joseph Wilson, and other individuals connected to the Program in Writing and Rhetoric.
35. Some of this history, including the crafting of this statement and early work we did around professional development, is documented in Sumyat Thu, Katie Malcolm, Candice Rai, and Anis Bawarshi, "Antiracist Translingual Praxis in Writing Ecologies," in *Writing Across Difference: Theory and Intervention*, ed. James Rushing Daniel, Katie Malcolm, and Candice Rai (Logan: Utah University Press, 2022), 195–217.
36. Caroline Gottschalk Druschke, "A Trophic Future for Rhetorical Ecologies," *Enculturation*, February 10, 2019.
37. Staff of Ron DeSantis, "Governor DeSantis Announces Legislative Proposal to Stop W.O.K.E. Activism and Critical Race Theory in Schools and Corporations," press release, December 15, 2021, Ron DeSantis, 46th Governor of Florida, official website, https://www.flgov.com/.
38. "CS/HB 7: Individual Freedom," general bill draft by Education and Employment Committee of the Florida House of Representatives, March 10, 2022.
39. "PWR Statement on Antiracist Writing Pedagogy and Program Praxis."
40. This is a composite story that combines several I experienced in our program. The details are kept spare and represent similar stories that occurred in multiple situations.
41. Jennifer Seibel Trainor, *Rethinking Racism: Emotion, Persuasion, and Literacy*

Education in an All-White High School (Carbondale: Southern Illinois University Press, 2008).

MICROCLIMATE FIVE

1. Walter D. Mignolo and Catherine E. Walsh, *On Decoloniality: Concepts, Analytics, Praxis* (Durham, NC: Duke University Press, 2018), 106.
2. Mignolo and Walsh, *On Decoloniality*, 100.
3. Shamard Charles, "What Is 'Weathering'? The Phenomenon that Is Killing Black People Slowly," *The Grio*, September 14, 2021; Max Liboiron, *Pollution Is Colonialism* (Durham, NC: Duke University Press, 2021).
4. Charles, "What Is 'Weathering?'"

CHAPTER SIX. **VIOLENT CLOUDS, ASHEN MEMORIES**

1. See the Clery Center web page for an overview, "At the Heart of Campus Safety." Cleary Center, 2022, https://www.clerycenter.org/. Andrea Stagg and Joseph Storch, "An Overview of the Violence Against Women Act (VAWA) Amendment to the Clery Act," (white paper, University Risk Management and Insurance Association Government and Regulatory Affairs, May 2017).
2. I shared this observation with our Office of Victim Assistance, and then the chief of campus police, who then shared my concern with police leadership across the Pac-12 network, leading to more tempered language on multiple campuses. Because these reports are by genre brief, they will never tell the full story.
3. Mitchel Byars, "Boulder Rape Renews Sentencing Outrage; Draws Comparisons to Brock Turner Case," *Denver Post*, August 11, 2016.
4. I do not wish to detract from studies of rhetorical, ecological violence. See Megan Eatman's writing on "constitutive" violence in *Ecological of Harm: Rhetorics of Violence in the United States* (Columbus: Ohio State University Press, 2020).
5. Gay Elwell, "Clery Was Alive during Dormitory Rape, Court Told," *Morning Call*, April 20, 1987.
6. Rob O'Dell and Anne Ryman, "'It Means Her Life Was Not in Vain': The Tragedy that Gave Birth to the Clery Act," *AZ Central*, April 15, 2016.
7. See Chanel Miller's *Know My Name: A Memoir* (New York: Penguin, 2019) for an

account of how "disposable memories" rigidly dictate a victim's life once violence obliterates the self, and the incident becomes a legal, mediated spectacle.

8. Lauren Berlant distinguishes cruel optimism from melancholy by insisting the former attaches desire to forever compromised conditions while melancholy invents a temporal framing for tragic loss, in *Cruel Optimism* (Durham, NC: Duke University Press, 2011), 24.
9. Charles, "What Is 'Weathering'?"
10. For example, consider Senator Mitch McConnell's whitewashing of any responsibility for anti-Black racism in the United States, as commented on by Ta-Nehisi Coates, "Reparations Are Not Just about Slavery but also Centuries of Theft & Racial Terror," *Democracy Now*, June 20, 2019.
11. Bryan Pietsch, "Melee near University of Colorado Boulder Injures 3 Officers," *New York Times*, March 7, 2021.
12. Andrew Kenney and Paola Zialcita, "10 Dead, Including a Police Officer, in Shooting at Boulder King Soopers," Colorado Public Radio News, March 22, 2021.
13. Daniel Victor and Derrick Bryson Taylor, "A Partial List of Mass Shootings in the United States in 2021," *New York Times*, December 1, 2021.
14. Dadja Popovich and Josh Katz, "See How Wildfire Smoke Spread across America," *New York Times*, July 21, 2021. Ankita Rao, Erum Salam, and Juweek Adolphe, "Which US States Make It Hardest to Vote?," *Guardian*, January 21, 2020. "Mass Shooting in America," *Everytown*, June 4, 2021. Frank Romo and Malcom MacLachlan. "Mapping the Black Lives Matter Movement." https://www.blm-map.com/. See Jacques Rancière, *The Future of the Image* (New York: Verso, 2009), 120. Rancière wants not only to elevate the unrepresentable to become the counterpoint to represented truth through art and speech but also to propose a "novelistic realism" as the "emancipation of semblance from representation."
15. Pluri-violence would be common to a "pluriverse" as a "world in which many worlds might fit" and to delink from worlds that the "West has managed to universalize" and which "only modern science can know and thoroughly study." Pluri-violence reveals itself to us as it does to the planet as "life as limitless flow." See Arturo Escobar, *Pluriversal Politics: The Real and the Possible* (Durham, NC: Duke University Press, 2020), 26.
16. The Rapid-Refresh modeling system, NOAA GAL, has been operating at NOAA's Global Systems Laboratory since July 2020. The reds, yellows, and purples of the actual Global Systems Laboratory smoke map can be found on the back cover of this book. The map is reproduced there with the gracious assistance of NOAA and the MSUP

team, and with the generosity of the reading group.

17. An argument in need of amplification, but for now, consider Alexander G. Weheliye's *Habeas Viscus: Racialized Assemblages, Biopolitics, and Black Feminist Theories of the Human* (Durham, NC: Duke University Press, 2014), especially chap. 4, "Racism: Biopolitics."
18. The best account of fire, heat, wind, climatic weirding, and noble efforts to track and predict a rogue calamity, and the painful story of losing a home to a shocking incapacity to do much more than flee—the best account I found is at the National Oceanic and Atmospheric Administration's "story map" of the Marshall Fire: ArcGIS StoryMaps, https://storymaps.arcgis.com/stories/cd7e211f5d594f9996b061d05670e779.
19. Tracy Brown and Lynn Berry, "The AP Interview: Fiona Hill Says Putin Has Host of Options," Associated Press, February 17, 2022.
20. Michelle Murphy, "Alterlife and Decolonial Chemical Relations," *Cultural Anthropology* 32, no. 4 (2017): 494–503.
21. Murphy, "Alterlife," 500.
22. John Durham Peters's thesis is that a "medium must not mean but be" and the "media are our infrastructures of being, the habitats and materials through which we act and are. This gives them ecological, ethical, and existential import. There is little as marvelous as the sea, the sky, or another person's presence, but most philosophy of media has rushed past these elements too quickly." John Durham Peters, *The Marvelous Clouds* (Chicago: University of Chicago Press, 2015), 13–15.
23. For a depiction of diagrammatical violence, see chap. 3, Brian Massumi, "The Political Economy of Belonging and the Logic of Relation," in *Parables for the Virtual* (Durham, NC: Duke University Press, 2002), 68–88.
24. Christina Sharpe, *In the Wake: On Blackness and Being* (Durham, NC: Duke University Press, 2016), 76.
25. Michelle Murphy, *The Economization of Life* (Durham, NC: Duke University Press, 2017), 134.
26. This essay embraces the assertion that rhetoric as a canon of study reinforces coloniality through procedural consensus-making, as outlined by Michael Lechuga in "An Anticolonial Future: Reassembling the Way We Do Rhetoric," *Communication and Critical/Cultural Studies* 17, no. 4: 378–85.
27. The police, after the King Soopers shootings, absolutely remember, as do trauma recovery units that wipe the blood from linoleum, cantaloupes, and metal shelving.
28. Foremost illustrated by Elie Wiesel in *Night*: "Never shall I forget that night, the

first night in camp, which has turned my life into one long night, seven times cursed, and seven times sealed. Never shall I forget that smoke. Never shall I forget the little faces of the children, whose bodies I saw turned into wreaths of smoke beneath a silent blue sky." Elie Wiesel, *Night*, trans. Marion Wiesel (New York: Hill and Wang, 2006), 34.

29. For an overview of societal forgetting, see Paul Connerton, "Seven Types of Forgetting," *Memory Studies* 1, no. 1 (2008): 59–71.
30. For more on "uneven" racial capitalism, see Katherine McKittrick, *Demonic Grounds: Black Women and the Cartographies of Struggle* (Minneapolis: University of Minnesota Press, 2006). For a clear account of how common settler technologies are to universities, see la paperson, *A Third University Is Possible* (Minneapolis: University of Minnesota, 2017).
31. Jason De León, *The Land of Open Graves: Living and Dying on the Migrant Trail* (Berkeley: University of California Press, 2015), 71.
32. Saidiya Hartman, *Lose Your Mother: A Journey along the Atlantic Slave Route* (New York: Macmillan, 2008), 6.
33. Hartman, *Lose Your Mother*, 133.
34. Sharpe, *In the Wake*, 18; Dionne Brand, "Verso 55," unpublished verse, used with permission, borrowed here.
35. Sharpe, *In the Wake*, 19.
36. Sharpe, *In the Wake*, 41.
37. Sharpe, *In the Wake*, 41.
38. De León, *The Land of Open Graves*, 69.
39. De León, *The Land of Open Graves*, 71.
40. De León, *The Land of Open Graves*, 72.
41. De León, *The Land of Open Graves*, 74.
42. De León, *The Land of Open Graves*, 76, 81.
43. Avishai Margalit, *The Ethics of Memory* (Cambridge, MA: Harvard University Press, 2004), 7.
44. Sharpe, *In the Wake*, 23.
45. Sharpe, *In the Wake*, 23.
46. Sharpe, *In the Wake*, 23.
47. Katherine McKittrick, "Plantation Futures," *Small Axe* 42 (November 2013): 2–3.
48. Murphy, *The Economization of Life*, 51.
49. Murphy, *The Economization of Life*, 52.
50. Murphy, *The Economization of Life*, 134.

51. Larry Buchanan, Quoctrung Bui, and Jugal K. Patel, "Black Lives Matter May Be the Largest Movement in U.S. History," *New York Times*, July 3, 2020.
52. See NASA, "Formation of Clouds Linked to Air Pollution," *Mongabay*, July 13, 2006. See Pierre Alexis Geoffroy and Ali Amad, "Seasonal Influence on Mass Shooting," *American Journal of Public Health* 106, no. 5 (2016). And then local police in Boulder comment that weather is not thought of as a tactical skill, but officers with experience prepare for upticks in domestic violence in the cooler months and street crime as climate warms. In the Climate Central report, "Tear Gas, Pollution, Wildfire Smoke: A Triple Threat to Your Lungs," PM 2.5 particulates enter all our lungs, but much more adversely for "low-income residents," and when Teron McGrew was asked about the authority to define public policy, "For 400 years, African Americans have been left out of the equation."
53. Juliana Spahr, "Poem Written after September 11," in *This Connection of Everyone with Lungs* (Berkeley: University of California Press, 2005).
54. In Lauren Berlant's "The Commons: Infrastructures for Troubling Times," *Environment and Planning D: Society and Space* (2016): 393–419. Spahr's oeuvre is featured as illustration of "transduction of the natural symbol into a revelation of ontological likeness" (401) and then in the collection, *This Connection of Everyone with Lungs*, where "Poem Written after September 11" can be found, of how "repair looks conventionally just like disrepair."
55. Glenn Albrecht, Gina-Maree Sartore, Linda Connor, Nick Higginbotham, Sonia Freeman, Brian Kelly, Helen Stain, Anne Tonna, and Georgia Pollard, "Solastalgia: The Distress Caused by Environmental Change," supplement, *Australasian Psychiatry* 15, no. 1 (2007): s95–s98.
56. Albrecht et al., "Solastalgia," 95–97.
57. Albrecht et al., "Solastalgia," 97.
58. Karen Bartlett, "These Men Offered 'Perfection in Cremation Technology' to the Nazis. We Can Learn from the Records They Left," *Time*, August 21, 2018.
59. Ernst Bornstein, *The Long Night: A True Story* (Jerusalem: Koren Publishers, 2015).
60. Hilary Swift and Corey Kilgannon, "9/11 Survivors Are Still Getting Sick Decades Later: 'Am I Next?'" *New York Times*, September 9, 2021.
61. Rachel Martin, "From Sacred Ground: A 9/11 Story," transcript from a September 5, 2021, National Public Radio, https://www.npr.org/transcripts/1034150099.
62. The image of this scene belongs to John Filo. My account of the economic aftermath discussed in John Ackerman, "Rhetorical Engagement in the Cultural Economies of Cities," in *The Public Work of Rhetoric: Citizen-Scholars and Civic Engagement*, ed.

John Ackerman and David Coogan (Columbia: University of South Carolina Press, 2010), 76–97.

63. Yael Navaro-Yashin, *The Make-Belief Space: Affective Geography in a Postwar Polity* (Durham, NC: Duke University Press, 2012), 150–51. I write about affective ruination in "Rhetorical Life among the Ruins," in *Field Rhetoric: Ethnography, Ecology, and Engagement in Place of Persuasions*, ed. Candice Rai and Caroline Gottschalk-Druschke (Tuscaloosa: University of Alabama Press, 2018), 171–92.
64. Charles Bowdin, *Murder City: Ciudad Juárez and the Global Economy's New Killing Fields* (New York: Bold Type, 2010), 22. My account is "'Biophilia': Works and Days."
65. Dionne Brand, *Ossuaries* (Toronto: McClelland & Stewart, 2010), 9–11.
66. Sharpe, *In the Wake*, 76, 109. She quotes unpublished verse by Dionne Brand, "Verso 55," used with permission, borrowed here.
67. Liboiron, *Pollution Is Colonialism*, 51.
68. Liboiron, *Pollution Is Colonialism*, 57.
69. Murphy, *The Economization of Life*, 134.

INCONCLUSION. **A READING GROUP MEETING ON *RHETORICAL CLIMATOLOGY***

1. Italo Calvino, *Invisible Cities*, trans. William Weaver (New York: Vintage, 1974), 165.

Works Cited

Ackerman, John. "Rhetorical Life among the Ruins." In *Field Rhetoric: Ethnography, Ecology, and Engagement in Place of Persuasions*, edited by Candice Rai and Caroline Gottschalk-Druschke, 171–92. Tuscaloosa: University of Alabama Press, 2018.

———. "Rhetorical Engagement in the Cultural Economies of Cities." In *The Public Work of Rhetoric: Citizen Scholars and Civic Engagement*, edited by John Ackerman and David Coogan, 76–97. Columbia: University of South Carolina Press, 2010.

Adam, Barbara. *Timescapes of Modernity: The Environment and Invisible Hazards*. London: Routledge, 2005.

Ahmed, Sara. *The Cultural Politics of Emotion*. 2nd ed. Edinburgh: Edinburgh University Press, 2014.

———. "Happy Objects." In *The Affect Theory Reader*, edited by Melissa Gregg and Gregory J. Seigworth, 29–51. Durham, NC: Duke University Press, 2010.

———. "Not in the Mood." *New Formations* 82 (2014): 13–28.

———. *On Being Included: Racism and Diversity in Institutional Life*. Durham, NC: Duke University Press, 2012.

———. *Willful Subjects*. Durham, NC: Duke University Press, 2014.

Alabama Department of Public Health. "Annex to ESF 8 of the State of Alabama Emergency Operations Plan," Alabama Department of Public Health 2010, https://www.alabamapublichealth.gov/ [page removed from the department's website in April 2020].

Albano, Catarina. *Out of Breath: Vulnerability of Air in Contemporary Art*. Minneapolis: University of Minnesota Press, 2022.

Albrecht, Glenn, Gina-Maree Sartore, Linda Connor, Nick Higginbotham, Sonia

Freeman, Brian Kelly, Helen Stain, Anne Tonna, and Georgia Pollard. "Solastalgia: The Distress Caused by Environmental Change." Supplement, *Australasian Psychiatry* 15, no. 1 (2007): s95–s98.

Amato, Joseph A. *Surfaces: A History*. Berkeley: University of California Press, 2013.

Appadurai, Arjun. *Fear of Small Numbers: An Essay on the Geography of Anger*. Durham, NC: Duke University Press, 2006.

Arendt, Hannah. *The Human Condition*. Chicago: University of Chicago Press, 1958.

Aristotle. *Politics*. Trans. C. D. C. Reeve. New York: Hackett Publishing, 1998.

Asante, Godfried Agyeman. "#RhetoricSoWhite and US Centered: Reflections on Challenges and Opportunities." *Quarterly Journal of Speech* 105, no. 4 (2019): 484–88.

Asante, Molefi Kete. *Afrocentricity: The Theory of Social Change*. Chicago: African American Images, 2003.

Baker-Bell, April. *Linguistic Justice: Black, Language, Literacy, Identity, and Pedagogy*. Urbana, IL: National Council of Teachers of English, 2020.

———. "*We Been Knowin*: Toward an Antiracist Language & Literacy Education." *Journal of Language and Literacy Education* 16, no. 1 (Spring 2020): 1–12.

Baldwin, James. "On Being White and Other Lies." In *The Cross of Redemption: Uncollected Writings*, edited by Randall Kenan, 135–38. New York: Pantheon, 2010.

Barad, Karen. *Meeting the Universe Halfway*. Durham, NC: Duke University Press, 2007.

Barbarin, Imani (@Imani_Barbarin). "I think about the time an abled random stranger threw my crutch into the pool 'to help me swim' a lot." Twitter, March 15, 2019, 6:14 p.m.

Barlow, Richard G. "Infinite Worlds: Robert Burton's Cosmic Voyage." *Journal of the History of Ideas* 34, no. 2 (1973): 291–302.

Barnett, Joshua Trey. *Mourning in the Anthropocene: Ecological Grief and Earthly Coexistence*. East Lansing: Michigan State University Press, 2022.

———. "Rhetoric for Earthly Coexistence: Imagining an Ecocentric Rhetoric." *Rhetoric & Public Affairs* 24, nos. 1–2 (2021): 365–78.

Bartlett, Karen. "These Men Offered 'Perfection in Cremation Technology' to the Nazis. We can Learn from the Records They Left." *Time*, August 21, 2018.

Bataille, Georges. *Visions of Excess: Selected Writings, 1927–1939*. Minneapolis: University of Minnesota Press, 1985.

Beardsley, John. "The Massachusetts Body of Liberties." General Court of the Commonwealth of Massachusetts Bay, December, 1641. Winthrop Society. Updated 2015. https://www.winthropsociety.com/.

Bennett, Jane. *Vibrant Matter*. Durham, NC: Duke University Press, 2010.

Berlant, Lauren. "The Commons: Infrastructures for Troubling Times." *Environment and Planning D: Society and Space* 34, no. 3 (2016): 393–419.

———. *Cruel Optimism*. Durham, NC: Duke University Press, 2011.

Bitzer, Lloyd F. "The Rhetorical Situation." *Philosophy & Rhetoric* 25 (1992): 1–14.

Blank, Radha, dir. *The Forty-Year Old Version*. Encino, CA: Hillman Grad Production, 2020.

Boia, Lucian. *The Weather in the Imagination*. London: Reaktion Books, 2005.

Bonilla-Silva, Eduardo. *Racism without Racist: Color-Blind Racism and the Persistence of Racial Inequality in American*. Lanham, MD: Rowman & Littlefield, 2013.

Booth, Wayne. *The Rhetoric of Rhetoric*. Malden, MA: Blackwell, 2004.

Bornstein, Ernst. *The Long Night: A True Story*. Jerusalem: Koren Publishers, 2015.

Bowden, Charles. *Murder City: Ciudad Juárez and the Global Economy's New Killing Fields*. New York: Nation, 2010.

Brand, Dionne. *The Blue Clerk: Ars Poetica in 50 Versos*. Durham, NC: Duke University Press, 2018.

———. *Ossuaries*. Toronto: McClelland and Stewart, 2010.

Brandzel, Amy. *Against Citizenship: The Violence of the Normative*. Urbana: University of Illinois Press, 2016.

Brathwaite, Edward Kamau, and Nathaniel Mackey. "An Interview with Edward Kamau Brathwaite." *Hambone* 9 (1991): 42–59.

Brauer, Ralph W. "Geography in the Medieval Muslim World: Seeking a Basis for Comparison of the Development of the Natural Sciences in Different Cultures." *Comparative Civilizations Review* 26, no. 5 (1992): 73–110.

Brown, Tracy, and Lynn Berry. "The AP Interview: Fiona Hill Says Putin Has Host of Options." Associated Press, February 17, 2022.

Buchanan, Larry, Quoctrung Bui, and Jugal K. Patel. "Black Lives Matter May Be the Largest Movement in U.S. History," *New York Times*, July 3, 2020.

Burke, Kenneth. *Permanence and Change: An Anatomy of Purpose*. 3rd ed. Berkeley: University of California Press, 1984.

Burton, Robert. *The Anatomy of Melancholy*. The Ex-Classics Project. 2009. https://www.exclassics.com.

Butler, Judith. *The Force of Nonviolence: An Ethico-Political Bind*. London: Verso, 2020.

Byars, Mitchel. "Boulder Rape Renews Sentencing Outrage; Draws Comparisons to Brock Turner Case." *Denver Post*, August 11, 2016.

Calvino, Italo. *Invisible Cities*. Translated by William Weaver. New York: Vintage, 1974.

Carpenter, Rick. "Disability as Socio-Rhetorical Action: Towards a Genre-Based Approach." *Disability Studies Quarterly* 31, no. 3 (2011).

Ceccarelli, Leah. *On the Frontier of Science: An American Rhetoric of Exploration and Exploitation*. East Lansing: Michigan State University Press, 2013.

———. "Polysemy: Multiple Meanings in Rhetorical Criticism." *Quarterly Journal of Speech* 84, no. 4 (1998): 395–415.

Cedillo, Christina V. "What Does It Mean to Move? Race, Disability, and Critical Embodiment Pedagogy." *Composition Forum* 39 (2018).

Césaire, Aimé. *Discourse on Colonialism*. Translated by Joan Pinkham. New York: Monthly Review Press, 2000.

Chakrabarty, Dipesh. "Anthropocene Time." *History and Theory* 57, no. 1 (2018): 5–32.

Chandler, Nahum Dmitri. *X—The Problem of the Negro as a Problem for Thought*. New York: Fordham University Press, 2014.

Charles, Shamard. "What Is 'Weathering'? The Phenomenon That Is Killing Black People Slowly." *The Grio*, September 14, 2021.

Chávez, Karma R. "Beyond Complicity: Coherence, Queer Theory, and the Rhetoric of the 'GayChristian Movement.'" *Text & Performance Quarterly* 24, nos. 3–4 (2004): 255–75.

Chen, Mel Y. *Animacies: Biopolitics, Racial Mattering, and Queer Affect.* Durham, NC: Duke University Press, 2012.

Cheng, Anne Anlin. *Ornamentalism*. New York: Oxford University Press, 2019.

Cherney, James L. *Ableist Rhetoric: How We Know, Value, and See Disability.* State College: Penn State University Press, 2019.

———. "The Rhetoric of Ableism." *Disability Studies Quarterly* 31, no. 3 (2011).

Cheyne, Ria. *Disability, Literature, Genre: Representation and Affect in Contemporary Fiction*. Liverpool: Liverpool University Press, 2019.

Chiu, Allyson. "A Fox News Guest Called Greta Thunberg 'Mentally Ill.' The Network Apologized for the 'Disgraceful' Comment." *Washington Post*, September 24, 2019.

Clare, Eli. *Exile and Pride: Disability, Queerness, and Liberation.* Durham, NC: Duke University Press, 2015.

Clery Center. "At the Heart of Campus Safety." Clery Center. 2022. www.clerycenter.org/.

Coates, Ta-Nehisi. "Reparations Are Not Just about Slavery but Also Centuries of Theft and Racial Terror." *Democracy Now*, June 20, 2019.

Coleman, Rebecca, and Liz Oakley-Brown. "Visualizing Surfaces, Surfacing Vision: Introduction." Special section, *Theory, Culture, and Society* 34, nos. 7–8 (2017): 5–108.

Condon, Frankie, and Vershawn Ashanti Young. "Introduction." In *Performing Antiracist Pedagogy in Rhetoric, Writing, and Communication*, edited by Frankie Condon and Vershawn Ashanti Young, 3–16. Boulder: WAC Clearinghouse, 2017.

Connell, Raewyn. *Southern Theory: The Global Dynamics of Knowledge in Social Science*.

Boston: Polity, 2007.

Connerton, Paul. "Seven Types of Forgetting." *Memory Studies* 1, no. 1 (2008): 59–71.

Cram, E. *Violent Inheritance: Sexuality, Land, and Energy in Making the North American West*. Oakland: University of California Press, 2022.

Crenshaw, Kimberlé. "Mapping the Margins: Intersectionality, Identity Politics, and Violence against Women." In *Critical Race Theory: The Key Writings that formed the Movement*, edited by Kimberlé Crenshaw, Neil Gotanda, Gary Pellar, and Kendall Thomas, 357–83. New York: New Press, 1996.

Crespo, Gisela. "This Store Employee's Simple Gesture Meant the World to a Teen with Autism." *CNN*, August 3, 2018. https://www.cnn.com.

Davis, Diane. *Inessential Solidarity: Rhetoric and Foreigner Relations*. Pittsburgh: University of Pittsburgh Press, 2010.

Davis, Olga Idriss. "A Black Woman as Rhetorical Critic: Validating Self and Violating the Space of Otherness." *Women's Studies in Communication* 21, no. 1 (1998): 77–89.

De León, Jason. *The Land of Open Graves: Living and Dying on the Migrant Trail*. Berkeley: University of California Press, 2015.

DeLuca, Kevin Michael. *Image Politics: The New Rhetoric of Environmental Activism*. New York: Routledge, 2005.

DeLuca, Kevin Michael, and Anne Teresa Demo. "Imaging Nature: Watkins, Yosemite, and the Birth of Environmentalism." *Critical Studies in Media Communication* 17, no. 3 (2000): 241–60.

Devitt, Amy J. *Writing Genres*. Carbondale: Southern Illinois University Press, 2004.

Dobrin, Sid. *Natural Discourse: Toward Ecocomposition*. Albany: State University of New York Press, 2002.

———. *Postcomposition*. Carbondale: Southern Illinois University Press, 2011.

Dolmage, Jay Timothy. *Disabled upon Arrival: Eugenics, Immigration, and the Construction of Race and Disability*. Columbus: Ohio State University Press, 2018.

———. *Disability Rhetoric*. Syracuse, NY: Syracuse University Press, 2014.

———. "Metis, *Mêtis, Mestiza*, Medusa: Rhetorical Bodies across Rhetorical Traditions." *Rhetoric Review* 28, no. 1 (2009): 1–28.

Doyle, Julie. *Mediating Climate Change*. New York: Routledge, 2011.

Du Bois, W. E. B. *The Souls of Black Folk*. G&D Media, 2019.

Eatman, Megan. *Ecologies of Harm*. Columbus: Ohio State University Press, 2020.

Edbauer, Jenny. "Unframing Models of Public Distribution: From Rhetorical Situation to Rhetorical Ecologies." *Rhetoric Society Quarterly* 35, no. 4 (2005): 5–24.

Ellis, Katie. *Disability and Popular Culture: Focusing Passion, Creating Community and*

Expressing Defiance. New York: Routledge, 2016.
Elwell, Gay. "Clery Was Alive during Dormitory Rape, Court Told." *Morning Call*, April 20, 1987.
Emmons, Kimberly. "Uptake and the Biomedical Subject." *Genre in a Changing World*, edited by Charles Bazerman, Adair Bonini, and Debora Figueiredo, 134–57. Anderson, SC: Parlor Press, 2009.
Erevelles, Nirmala. *Disability and Difference in Global Contexts: Enabling a Transformative Body Politic*. New York: Palgrave Macmillan, 2011.
Escobar, Arturo. *Pluriversal Politics: The Real and the Possible*. Durham, NC: Duke University Press, 2020.
Fanon, Frantz. *Black Skin, White Masks*. Translated by Charles Lam Markmann. London: Pluto Press, 1986.
Flores, Lisa A. "Between Abundance and Marginalization: The Imperative of Racial Rhetorical Criticism." *Review of Communication* 6, no. 1 (2015): 4–24.
———. *Deportable and Disposable: Public Rhetoric and the Making of the "Illegal" Immigrant*. University Park: Penn State University Press, 2020.
Flores, Lisa A., and Mary Ann Villareal. "Mobilizing for National Inclusion: The Discursivity of Whiteness among Texas Mexicans' Arguments for Desegregation." In *Border Rhetorics: Citizenship and Identity on the US–Mexico Frontier*, edited by D. Robert DeChaine, 86–100. Tuscaloosa: University of Alabama, 2012.
Frey, Lawrence, and Joshua Hanan. "Toward Social Justice Activism Critical Rhetoric Scholarship." *International Journal of Communication* 14 (2020): 850–69.
Friedner, Michelle, and Karen Weingarten. "Introduction: Disorienting Disability." *South Atlantic Quarterly* 118, no. 3 (2019): 483–90.
Garland-Thomson, Rosemarie. *Extraordinary Bodies: Figuring Physical Disability in American Culture and Literature*. New York: Columbia University Press, 1997.
———. "Seeing the Disabled: Visual Rhetorics of Disability in Popular Photography." In *The New Disability History*, edited by Paul K. Longmore and Lauri Umansky, 335–74. New York: NYU Press, 2001.
Geoffroy, Pierre Alexis, and Ali Amad. "Seasonal Influence on Mass Shooting." *American Journal of Public Health* 106, no. 5 (2016).
Glissant, Édouard. *The Poetics of Relation*. Translated by Betsy Wing. Ann Arbor: University of Michigan Press, 2010.
Gordon, Lewis R. "Race, Theodicy, and the Normative Emancipatory Challenges of Blackness." *South Atlantic Quarterly* 112, no. 4 (2013): 725–36.
Gorman, Rachel. "Disablement In and For Itself: Toward a 'Global' Idea of Disability."

Somatechnics 6, no. 2 (2016): 249–61.

Gorsevski, Ellen W. *Peaceful Persuasion: The Geopolitics of Nonviolent Rhetoric*. Albany: State University of New York Press, 2004.

Gottschalk Druschke, Caroline. "A Trophic Future for Rhetorical Ecologies." *Enculturation* (February 10, 2019).

Gries, Laurie. *Still Life with Rhetoric: A New Materialist Approach for Visual Rhetorics*. Logan: Utah State University Press, 2015.

Grue, Jan. "The Problem with Inspiration Porn: A Tentative Definition and a Provisional Critique." *Disability & Society* 31, no. 6 (2016): 838–49.

Hartman, Saidiya V. *Lose Your Mother: A Journey Along the Atlantic Slave Route*. New York: Macmillan, 2008.

———. *Scenes of Subjection: Terror, Slavery, and Self-Making in Nineteenth-Century America*. New York: Oxford University Press, 1997.

———. *Wayward Lives, Beautiful Experiments: Intimate Histories of Social Upheaval*. New York: W. W. Norton, 2019.

Hartnett, Stephen. "Communication, Social Justice, and Joyful Commitment." *Western Journal of Communication* 74, no. 1 (2010): 68–93.

Hashlamon, Yanar. "Rhetoricity at the End of the World: Defining Rhetorical Debility under Neoliberal Capitalism." *Rhetoric Society Quarterly* 52, no. 1 (2022): 18–31.

Hauser, Gerard. "Teaching Rhetoric: Or Why Rhetoric Isn't Just Another Kind of Philosophy or Literary Criticism." *Rhetoric Society Quarterly* 34, no. 3 (2004): 39–53.

Hawhee, Debra. *Rhetoric in Tooth and Claw: Animals, Language, Sensation*. Chicago: University of Chicago Press, 2017.

Hill, Charles A. "The Psychology of Rhetorical Images." In *Defining Visual Rhetorics*, edited by Charles A. Hill and Marguerite Helmers, 25–40. Mahwah, NJ: Lawrence Erlbaum, 2004.

Hippocrates. *Hippocrates on Airs, Water, and Places*. London: Wyman & Sons, 1881.

hooks, bell. *Black Looks: Race and Representation*. Boston, MA: South End Press, 1992.

Houdek, Matthew, and Ursula J. Ore. "Cultivating Otherwise Worlds and Breathable Futures." *Rhetoric, Politics & Culture* 1, no. 1, Summer 2021: 85–95.

Hulme, Mike. "Climate." *Environmental Humanities* 6 (2015): 175–78.

———. "Reducing the Future to Climate: A Story of Climate Determinism and Reductionism." *OSIRIS* 26 (2011): 245–66.

Ibn Khaldûn. *The Muqaddimah: An Introduction to History*. Translated by Franz Rosenthal. Princeton: Princeton University Press, 1967.

Ingold, Tim. *Lines: A Brief History*. London: Routledge, 2007.

———. "That's Enough about Ethnography!" *HAU: Journal of Ethnographic Theory* 4, no. 1 (2014): 383–95.

Ingraham, Chris. *Gestures of Concern*. Durham, NC: Duke University Press, 2020.

Inoue, Asao. *Antiracist Writing Assessment Ecologies: Teaching and Assessing Writing for a Socially Just Future*. Boulder: WAC Clearinghouse, 2015.

Jackson, Zakiyyah Iman. *Becoming Human: Matter and Meaning in an Antiblack World*. New York: NYU Press, 2020.

James, William. *Pragmatism: A New Name for Some Old Ways of Thinking*. New York: Longman, Green, 1907.

Jasanoff, Sheila. "Image and Imagination: The Formation of Global Environmental Consciousness." In *Science and Public Reason*, 78–102. London: Taylor & Francis Group, 2012.

———. "Ordering Knowledge, Ordering Society." In *States of Knowledge: The Co-Production of Science and Social Order*, edited by Shelia Jasanoff, 13–45. New York: Routledge, 2004.

Jasinski, James. *Sourcebook on Rhetoric: Key Concepts in Contemporary Rhetorical Studies*. Los Angeles: SAGE, 2001.

Juster, Norton. *The Phantom Tollbooth*. New York: Yearling Books, 1961.

Kafer, Alison. *Feminist, Queer, Crip*. Bloomington: Indiana University Press, 2013.

Keeling, Diane M., and Jennifer C. Prairie. "Trophic and Tropic Dynamics: An Ecological Perspective on Tropes." In *Tracing Rhetoric and Material Life: Ecological Approaches*, edited by Bridie McGreavy, Justine Wells, George F. McHendry Jr., and Samantha Senda-Cook, 39–58. New York: Palgrave Macmillan, 2018.

Keeling, Kara. *Queer Times, Black Futures*. New York: NYU Press, 2019.

Keith, William, Steve Fuller, Alan Gross, and Michael Leff. "Taking Up the Challenge: A Response to Simons." *Quarterly Journal of Speech* 85 (1999): 330–38.

Kelsie, Amber E. "Blackened Debate at the End of the World." *Philosophy and Rhetoric* 52, no. 1 (2019): 63–70.

Kennedy, Tammie M., Joyce Irene Middleton, and Krista Ratcliffe. *Rhetorics of Whiteness: Postracial Hauntings in Popular Culture, Social Media, and Education*. Carbondale: Southern Illinois University Press, 2017.

Kenner, Alison. *Asthma Care in a Time of Climate Change*. Minneapolis: University of Minnesota Press, 2018.

Kenney, Andrew, and Paola Zialcita. "10 Dead, Including a Police Officer, in Shooting at Boulder King Soopers." Colorado Public Radio News, March 22, 2021.

Kerschbaum, Stephanie L. *Toward a New Rhetoric of Difference*. Urbana, IL: National

Council of Teachers of English, 2014.

King, Tiffany Lethabo. *The Black Shoals: Offshore Formations of Black and Native Studies*. Durham, NC: Duke University Press, 2019.

Kjeldsen, Jens E. "Rhetoric as Working Through." Paper presented at the Norwegian Media Researchers Conference in Bergen, Norway, October 20–21, 2016.

———. "Talking to the Eye: Visuality in Ancient Rhetoric." *Word & Image* 19, no. 3 (July–September 2003): 133–37.

Knox-Russell, Allyse. "Futurity without Optimism: Detaching from Anthropocentrism and Grieving Our Fathers in *Beasts of the Southern Wild*." In *Affective Ecocriticism: Emotion, Embodiment, Environment*, edited by Kyle Bladow and Jennifer Ladino, 213–33. Lincoln: University of Nebraska Press, 2018.

Kuppers, Petra. "Toward a Rhizomatic Model of Disability: Poetry, Performance, and Touch." *Journal of Literary and Cultural Disability Studies* 3, no. 3 (2009): 221–40.

la paperson. *A Third University is Possible*. Minneapolis: University of Minnesota Press, 2017.

Lechuga, Michael. "An Anticolonial Future: Reassembling the Way We Do Rhetoric." *Communication and Critical/Cultural Studies* 17, no. 4 (2020): 378–85.

Lee, JongHwa, and Seth Kahn, eds. *Activism and Rhetoric*. New York: Routledge, 2020.

LeMesurier, Jennifer Lin. "Searching for Unseen Metic Labor in the Pussyhat Project." *Peitho* 22, no. 1 (Fall/Winter 2019).

———. "Uptaking Race: Genre, MSG, and Chinese Dinner." *POROI* 12, no. 2 (February 2017): 1–23.

Lewiecki-Wilson, Cynthia, and James C. Wilson. *Embodied Rhetorics: Disability in Language and Culture*. Carbondale: Southern Illinois University Press, 2001.

Lewis, Tanya. "The 'Shared Psychosis' of Donald Trump and His Loyalists." *Scientific American*, January 11, 2021.

Liboiron, Max. *Pollution Is Colonialism*. Durham, NC: Duke University Press, 2021.

Lindemann, Kurt, and James L. Cherney. "Communicating in and through 'Murderball': Masculinity and Disability in Wheelchair Rugby." *Western Journal of Communication* 72, no. 2 (April 2008): 107–25.

Linton, Simi. *Claiming Disability: Knowledge and Identity*. New York: NYU Press, 1998.

Lippi-Green, Rosina. *English with an Accent: Language, Ideology and Discrimination in the United States*. New York: Routledge, 2012.

Lorde, Audre. "The Master's Tools Will Never Dismantle the Master's House." In *Sister Outsider: Essays and Speeches*, 110–13. Berkeley: Crossing Press, 1984.

Luciano, Dana, and Mel Y. Chen. "Has the Queer Ever Been Human?" *GLQ* 21, no. 2–3

(2015): 183–207.

Lyons, Scott Richard. "Rhetorical Sovereignty: What Do American Indians Want from Writing?" *College Composition and Communication* 51, no. 3 (2000): 447–68.

MacLeish, Archibald. "Riders on Earth Together, Brothers in Eternal Cold." *New York Times*. December 25, 1968.

Macnaghten, Phil, and John Urry. *Contested Natures*. London: SAGE, 1998.

Margalit, Avishai. *The Ethics of Memory*. Cambridge, MA: Harvard University Press, 2004.

Marks, Wizard. "I-35W Disrupted Minority Community, Boxed-In Phillips." *The Alley*, August 1990.

Martin, Rachel. "From Sacred Ground: A 9/11 Story." Transcript from a September 5, 2021, National Public Radio. https://www.npr.org/transcripts/1034150099.

"Mass Shooting in America." *Everytown*, June 4, 2021.

Massumi, Brian. *Parables for the Virtual*. Durham, NC: Duke University Press, 2002.

Mbembe, Achille. *Critique of Black Reason*. Translated by Laurent Dubois. Durham, NC: Duke University Press, 2017.

McGee, Michael Calvin. "Text, Context, and the Fragmentation of Contemporary Culture." *Western Journal of Communication* 54 (1990): 274–89.

McGreavy, Bridie, Darren Ranco, John Daigle, Suzanne Greenlaw, Nolan Altvater, Tyler Quiring, Natalie Michelle, Jan Paul, Maliyan Binette, Brawley Benson, Anthony Sutton, and David Hart, "Science in Indigenous Homelands: Addressing Power and Justice in Sustainability Science from/with/in the Penobscot River." *Sustainability Science* 16, no. 3 (2021): 937–47.

McGreavy, Bridie, Sara Randall, Tyler Quiring, Carter Hathaway, and Gabrielle Hillyer. "Enhancing Adaptive Capacities in Coastal Communities through Engaged Communication Research: Insights from a Statewide Study of Shellfish Co-management." *Ocean & Coastal Management* 163 (2018): 240–53.

McKinnon, Sara L., Robert Asen, Karma R. Chávez, and Robert Glenn Howard, eds. *Text + Field: Innovations in Rhetorical Method*. University Park: Penn State University Press, 2016.

McKittrick, Katherine. *Demonic Grounds: Black Women and the Cartographies of Struggle*. Minneapolis: University of Minnesota Press, 2006.

———. "Plantation Futures." *Small Axe* 42 (November 2013): 1–15.

McLuhan, Marshall. "At the Moment of Sputnik the Planet Became a Global Theater in Which There Are No Spectators but Only Actors." *Journal of Communication* (Winter 1974): 48–58.

McPhail, Mark Lawrence. "A Question of Character: Re(-)signing the Racial Contract."

Rhetoric & Public Affairs 7, no. 3 (2004): 391–405.

Mercieca, Jennifer R. *Demagogue for President: The Rhetorical Genius of Donald Trump*. College Station: Texas A&M University Press, 2020.

Middleton, Michael, Aaron Hess, Danielle Endres, and Samantha Senda-Cook. *Participatory Critical Rhetoric: Theoretical and Methodological Foundations for Studying Rhetoric in Situ*. Lanham, MD: Lexington Books, 2015.

Mignolo, Walter. "Delinking." *Cultural Studies* 21, nos. 2–3 (2007): 449–514.

Miller, Carolyn R., Amy J. Devitt, and Victoria J. Gallagher. "Genre: Permanence and Change." *Rhetoric Society Quarterly* 38, no. 3 (2018): 269–77.

Miller, Chanel. *Know My Name: A Memoir*. New York: Penguin, 2019.

Milstein, Tema. "The Performer Metaphor: 'Mother Nature Never Gives Us the Same Show Twice.'" *Environmental Communication* 10, no. 2 (2016): 227–48.

Miquel, André. *La Géographie Humaine du Monde Musulman Jusqu'au Milieu du XI Siècle*. Paris: Mouton & Co., 1975.

Morrison, Aimée. "(Un)reasonable, (Un)necessary, and (In)appropriate: Biographic Mediation of Neurodivergence in Academic Accommodations." *Biography* 42, no. 2 (2019): 693–719.

Morton, Timothy. *The Ecological Thought*. Cambridge, MA: Harvard University Press, 2010.

Moten, Fred. *Black and Blur*. Durham, NC: Duke University Press, 2017.

———. *In the Break: The Aesthetics of the Black Radical Tradition*. Minneapolis: University of Minnesota Press, 2003.

———. *Stolen Life*. Durham, NC: Duke University Press, 2018.

———. *Universal Machine*. Durham, NC: Duke University Press, 2018.

Motha, Suhanthie. *Race, Empire, and English Language Teaching: Creating Responsible and Ethical Anti-Racist Practice*. New York: Teachers College Press, 2014.

Murphy, Michelle. "Against Population, Toward Alterlife." In *Making Kin, Not Population*, edited by Adele E. Clarke and Donna Jeanne Haraway, 101–24. Chicago: Pricky Paradigm Press, 2018.

———. "Alterlife and Decolonial Chemical Relations." *Cultural Anthropology* 32, no. 4 (2017): 494–503.

———. *The Economization of Life*. Durham, NC: Duke University Press, 2017.

Nakayama, Thomas K. "Whiteness Is Not Contained." *Communication and Critical/Cultural Studies* 17, no. 2 (2020): 199–201.

Nakayama, Thomas K., and Robert L. Krizek. "Whiteness: A Strategic Rhetoric." *Quarterly Journal of Speech* 81 (1995): 291–309.

Na'puti, Tiara R. "Archipelagic Rhetoric: Remapping the Marianas and Challenging Militarization from 'A Stirring Place.'" *Communication and Critical/Cultural Studies* 16, no. 1 (2019): 4–25.

NASA. "Formation of Clouds Linked to Air Pollution." *Mongabay*, July 13, 2006.

———. "NASA—What's the Difference between Weather and Climate?" NASA, February 1, 2005. https://www.nasa.gov.

National Oceanic and Atmospheric Administration. "RAP-Smoke, RAP-NCEP 07/2/2021." Global Systems Laboratory. https://rapidrefresh.noaa.gov/RAPsmoke/displayMapLocalDiskDateDomainZipTZA.cgi?keys=rap_ncep_smoke_jet:&runtime=2021072206&plot_type=vis_sfc&fcst=11&time_inc=60&num_times=52&model=rr&ptitle=RAP-Smoke%20Model%20Fields&maxFcstLen=51&fcstStrLen=-1&domain=full&adtfn=1.

———. "'Story Map' of the Marshall Fire." ArcGIS StoryMaps. https://storymaps.arcgis.com/stories/cd7e211f5d594f9996b061d05670e779.

National Resource Defense Council. "Flint Water Crisis: Everything You Need to Know." National Resource Defense Council, November 8, 2018. https://www.nrdc.org/stories/flint-water-crisis-everything-you-need-know.

Navaro-Yashin, Yael. "Affective Spaces, Melancholic Objects, Ruination and the Production of Anthropological Knowledge." *Journal of the Royal Anthropological Institute* 15 (2009): 1–18.

———. *The Make-Believe Space: Affective Geography in a Postwar Polity*. Durham, NC: Duke University Press, 2012.

Newman, Sara J. "Disability and Life Writing: Reports from the Nineteenth-Century Asylum." *Journal of Literary & Cultural Disability Studies* 5, no. 3 (2011): 261–78.

Nishime, Leilani, and Kim D. Hester Williams. *Racial Ecologies*. Seattle: University of Washington Press, 2018.

Nixon, Rob. "The Anthropocene: The Promise and Pitfalls of an Epochal Idea." In *Future Remains: A Cabinet of Curiosities for the Anthropocene*, edited by Gregg Mittman, Marco Armiero, and Robert Emmett, 1–18. Chicago: University of Chicago Press, 2018.

Nyong'o, Tavia. *Afro-Fabulations: The Queer Drama of Black Life*. New York: NYU Press, 2019.

O'Dell, Rob, and Anne Ryman. "'It Means Her Life Was Not in Vain': The Tragedy That Gave Birth to the Clery Act." *AZ Central*, April 15, 2016.

Omi, Michael, and Howard Winant. *Racial Formations in the United States: From the 1960s to the 1990s*. 2nd ed. New York: Routledge, 1994.

Ore, Ersula J. *Lynching: Violence, Rhetoric, and American Identity*. Jackson: University Press of Mississippi, 2019.

Ortiz, Naomi. "Majestic Disabled/Queer/People of Color Elders Instruct How to Dance in the Struggle." EcoSemantics Archive: Field Notes from the 2020 EcoSomatics Symposium. *Jacket2*, May 6, 2022.

Parker, Claire. "The Protests and Unrest That Defined 2021: Coronavirus, Climate Change and the Capital Riots." *Washington Post*, December 21, 2021.

Perryman-Clark, Staci M., and Collin Craig. "Black Matters: Writing Program Administration in Twenty-First-Century Higher Education." In *Black Perspectives in Writing Program Administration: From the Margins to the Center*, edited by Staci M Perryman-Clark and Collin Craig, 1–27. Urbana, IL: National Council of Teachers of English, 2019.

Peters, John Durham. *The Marvelous Clouds*. Chicago: University of Chicago Press, 2015.

Pietsch, Bryan. "Melee near University of Colorado Boulder Injures 3 Officers." *New York Times*, March 7, 2021.

Pinsky, Robert. "An Explanation of America." In *The Figured Wheel*. New York: Farrar, Straus and Giroux, 1997.

Plato. *Timaeus and Critias*. Translated by Desmond Lee. New York: Penguin Classics, 2008.

Plumwood, Val. *Feminism and the Mastery of Nature*. New York: Routledge, 1993.

Poole, Robert. *Earthrise: How Man First Saw the Earth*. New Haven, CT: Yale University Press, 2010.

Popovich, Dadja, and Josh Katz. "See How Wildfire Smoke Spread across America." *New York Times*, July 21, 2021.

Prelli, Lawrence J., Floyd D. Anderson, and Matthew T. Althouse. "Kenneth Burke on Recalcitrance." *Rhetoric Society Quarterly* 41, no. 2 (2011): 97–124.

Prendergast, Catherine. *Literacy and Racial Justice: The Politics of Learning after Brown v. Board of Education*. Carbondale: Southern Illinois University Press, 2003.

Pryal, Katie R. G. "The Genre of the Mood Memoir and the Ethos of Psychiatric Disability." *Rhetoric Society Quarterly* 40, no. 5 (2010): 479–501.

Puar, Jasbir K. *The Right to Maim: Debility, Capacity, Disability*. Durham, NC: Duke University Press, 2017.

Pugilese, Joseph. "Biometrics, Infrastructural Whiteness, and the Racialized Zero Degree of Nonrepresentation." *boundary 2* 34, no. 2 (2007): 105–33.

"PWR Statement on Antiracist Writing Pedagogy and Program Praxis." Program in Writing and Rhetoric, University of Washington. 2018, revised 2021. https://english.

washington.edu/diversity-equity-and-justice.

Rai, Candice, and Caroline Gottschalk Druschke, eds. *Field Rhetoric: Ethnography, Ecology, and Engagement in the Places of Persuasion*. Tuscaloosa: University of Alabama Press, 2018.

Rancière, Jacques. *The Future of the Image*. New York: Verso, 2009.

Rao, Ankita, Erum Salam, and Juweek Adolphe. "Which US States Make It Hardest to Vote?" *Guardian*, January 21, 2020.

Ratcliff, Krista. *Rhetorical Listening: Identification, Gender, Whiteness*. Carbondale: Southern Illinois University Press, 2006.

Ray, Rashawn. "Why Are Blacks Dying at Higher Rates from COVID-19?" *Brookings*, April 9, 2020.

Rice, Jenny Edbauer. "The New 'New': Making a Case for Critical Affect Studies." *Quarterly Journal of Speech* 94, no. 2 (2008): 200–212.

Rickert, Thomas. *Ambient Rhetoric: The Attunements of Rhetorical Being*. Pittsburgh: University of Pittsburgh Press, 2013.

Romo, Frank, and Malcom MacLachlan. "Mapping the Black Lives Matter Movement." https://www.blm-map.com/.

Roose, Kevin. "Sorry, but Working from Home Is Overrated." *New York Times*, March 10, 2020.

Rowland, Allison. *Zoetropes and the Politics of Humanhood*. Columbus: Ohio State University Press, 2020.

Sample, Ian. "*Earthrise*: How the Iconic Image Changed the World." *Guardian*, December 24, 2018.

Sampson, Tony. *Virality: Contagion Theory in the Age of Networks*. Minneapolis: University of Minnesota Press, 2012.

Sanderson, Marie. "The Classification of Climates from Pythagoras to Koeppen." *Bulletin of the American Meteorological Society* (1999): 669–73.

Sauzet, Sophie. "Phenomena—Agential Realism." *COST Action IS1307 New Materialism—Networking European Scholarship on 'How Matter Comes to Matter,'* March 13, 2018.

Scappetone, Jennifer. "Precarity Shared: Breathing as Tactic in Air's Uneven Commons." In *Precarity Shared: Breathing as Tactic in Air's Uneven Commons*, edited by Myung Mi Kim and Cristanne Miller, 41–57. Albany: State University of New York Press, 2018.

Schalk, Sami. "Black Disability Gone Viral: A Critical Race Approach to Inspiration Porn." *College Language Association Journal* 64, no. 1 (2021): 100–119.

———. "Reevaluating the Supercrip." *Journal of Literary & Cultural Disability Studies* 10, no. 1 (2016): 71–86.

Schalk, Sami (@DrSamiSchalk). "This discursive approach is ableist bc it suggests that people with psych disabilities are inherently dangerous, dismissible, disposable, etc. It adds to the oppression of folks with psych disabilities, relies on this oppression to be effective." Twitter, April 25, 2020, 12:33 p.m.

Schiappa, Edward. "Second Thoughts on the Critiques of Big Rhetoric." *Philosophy and Rhetoric* 34, no. 3 (2001): 260–74.

Schwarz, Astrid, and Kurt Jax. "Etymology and Original Sources of the Term 'Ecology.'" In *Ecology Revisited: Reflecting on Concepts, Advancing Science*, edited by Astrid Schwartz and Kurt Jax, 144–47. New York: Springer, 2011.

Segal, Judy Z. *Health and the Rhetoric of Medicine*. Carbondale: Southern Illinois University Press, 2008.

Sexton, Jared. "People-of-Color-Blindness: Notes on the Afterlife of Slavery." *Social Text* 28, no. 2 (103) (2010): 31–56.

Sharpe, Christina. *In the Wake: On Blackness and Being*. Durham, NC: Duke University Press, 2016.

Silva, Denise Farreira da. "Toward a Black Feminist Poethics: The Quest(ion) of Blackness Toward the End of the World." *Black Scholar* 44, no. 2 (2014): 81–97.

———. *Toward a Global Idea of Race*. Minneapolis: University of Minnesota Press, 2007.

Skinnell, Ryan. "Using Democracy against Itself: Demagogic Rhetoric as an Attack on Democratic Institutions." *Rhetoric Society Quarterly* 49, no. 3 (2019): 248–63.

Sloterdijk, Peter. *Terror from the Air*. Translated by Amy Patton and Steve Corcoran. Cambridge, MA: MIT Press, 2009.

Smith, Woodruff D. "Friedrich Ratzel and the Origins of Lebensraum." *German Studies Review* 3, no. 1 (1980): 51–68.

Spahr, Juliana. *This Connection of Everyone with Lungs*. Berkeley: University of California Press, 2005.

Spillers, Hortense J. *Black, White and in Color: Essays on American Literature and Culture*. Chicago: University of Chicago Press, 2003.

———. "Mama's Baby, Papa's Maybe: An American Grammar Book." In *The Black Feminist Reader*, edited by Joy James and T. Denean Sharpley-Whiting, 57–87. Malden, MA: Blackwell, 2000.

Staff of Ron DeSantis. "Governor DeSantis Announces Legislative Proposal to Stop W.O.K.E. Activism and Critical Race Theory in Schools and Corporations." Press release, December 15, 2021. Ron DeSantis, 46th Governor of Florida, official website.

https://www.flgov.com/.

Stagg, Andrea, and Joseph Storch. "An Overview of the Violence against Women Act (VAWA) Amendment to the Clery Act." White paper, University Risk Management and Insurance Association Government and Regulatory Affairs, May 2017.

Steigerwald, Bill. "NASA Releases New High-Resolution *Earthrise* Image." NASA. Updated August 7, 2017. https://www.nasa.gov/image-feature/goddard/lro-earthrise-2015.

Stormer, Nathan. "Articulation: A Working Paper on Rhetoric and Taxis." *Quarterly Journal of Speech* 90, no. 3 (2004): 257–84.

———. "Rhetoric's Diverse Materiality: Polythetic Ontology and Genealogy." *Review of Communication* 16, no. 4 (2016): 299–316.

Stormer, Nathan, and Bridie McGreavy. "Thinking Ecologically about Rhetoric's Ontology: Capacity, Vulnerability, and Resilience." *Philosophy and Rhetoric* 50, no. 1 (2017): 1–25.

Swift, Hilary, and Corey Kilgannon. "9/11 Survivors Are Still Getting Sick Decades Later: 'Am I Next?'" *New York Times*, September 9, 2021.

TallBear, Kim. *Native American DNA: Tribal Belonging and the False Promise of Genetic Science*. Minneapolis: University of Minnesota Press, 2013.

———. "Standing with and Speaking as Faith: A Feminist-Indigenous Approach to Inquiry." *Journal of Research Practice* 10, no. 2 (2014).

Taylor, Paul C. *Black Is Beautiful: A Philosophy of Black Aesthetics*. Hoboken, NJ: Wiley, 2016.

Thu, Sumyat, Katie Malcolm, Candice Rai, and Anis Bawarshi. "Antiracist Translingual Praxis in Writing Ecologies." In *Writing across Difference: Theory and Intervention*, edited by James Rushing Daniel, Katie Malcolm, and Candice Rai, 195–217. Logan: Utah University Press, 2022.

Towns, Armond. "Black 'Matter' Lives." *Women's Studies in Communication* 41, no. 4 (2018): 349–58.

———. "Rebels of the Underground: Media, Orality, and the Routes of Black Emancipation." *Communication and Critical/Cultural Studies* 13, no. 2 (2016): 184–97.

Trainor, Jennifer Seibel. *Rethinking Racism: Emotion, Persuasion, and Literacy Education in an All-White High School*. Carbondale: Southern Illinois University Press, 2008.

Tsing, Anna Lowenhaupt. *The Mushroom at the End of the World*. Princeton: Princeton University Press, 2015.

Tsing, Anna Lowenhaupt, Andrew S. Mathews, and Nils Bubandt. "Patchy Anthropocene: Landscape Structure, Multispecies History, and the Retooling of

Anthropology." *Current Anthropology* 60, no. 20 (2019): S186–S197.

Tuck, Eve, and K. Wayne Yang. "Decolonization Is Not a Metaphor." *Decolonization: Indigeneity, Education & Society* 1, no. 1 (2012): 1–40.

Tyler, Stephen, and Ivo Strecker. "The Rhetoric Culture Project." In *Culture + Rhetoric*, edited by Stephen Tyler and Ivo Strecker. New York: Berghahn Books, 2009.

Upton, John, and Danielle Venton. "Tear Gas, Pollution, Wildfire Smoke: A Triple Threat to Your Lungs." *Climate Central*, June 11, 2020.

Van Denburg, Hart. "PHOTOS: Aerial View of Neighborhoods Shows Devastation in Boulder County." Colorado Public Radio News, December 31, 2021.

Victor, Daniel, and Derrick Bryson Taylor. "A Partial List of Mass Shootings in the United States in 2021." *New York Times*, December 1, 2021.

Vidali, Amy. "Texts of Our Institutional Lives: Performing the Rhetorical Freak Show; Disability, Student Writing, and College Admissions." *College English* 69, no. 6 (2007): 615–41.

Walters, Shannon. *Rhetorical Touch: Disability, Identification, Haptics*. Columbia: University of South Carolina Press, 2014.

Wanzer-Serrano, Darrell. "Rhetoric's Rac(e/ist) Problems." *Quarterly Journal of Speech* 105, no. 4 (2019): 465–76.

Warren, Calvin L. *Ontological Terror: Blackness, Nihilism, and Emancipation*. Durham, NC: Duke University Press, 2018.

Washington, Harriet A. *Medical Apartheid: The Dark History of Medical Experimentation on Black Americans from Colonial Times to the Present*. New York: Harlem Moon, 2008.

Watts, Eric King. "Critical Cosmopolitanism, Antagonism, and Social Suffering." *Quarterly Journal of Speech* 101, no. 1 (2015): 271–79.

———. "Postracial Fantasies, Blackness, and Zombies." *Communication and Critical/Cultural Studies* 14, no. 4 (2017): 317–33.

Weheliye, Alexander G. *Habeas Viscus: Racializing Assemblages, Biopolitics, and Black Feminist Theories of the Human*. Durham, NC: Duke University Press, 2014.

Whyte, Kyle Powys. "Against Crisis Epistemology." In *Handbook of Critical Indigenous Studies*, edited by Brendan Hokowhitu, Aileen Moreton-Robinson, Linda Tuhiwai-Smith, Chris Andersen, and Steve Larkin, 52–64. New York: Routledge, 2021.

Wiesel, Elie. *Night*. Translated by Marion Wiesel. New York: Hill and Wang, 2006.

Wilderson, Frank B., III. *Afropessimism*. New York: W.W. Norton, 2020.

———. *Red, White & Black: Cinema and the Structure of U.S. Antagonisms*. Durham, NC: Duke University Press, 2010.

Williams, David R. "The Apollo 8 Christmas Eve Broadcast." NASA Goddard Space

Flight Center. Updated September 25, 2007. https://nssdc.gsfc.nasa.gov/planetary/lunar/apollo8_xmas.html.

Wilson, Kirt H. 1999. "Towards a Discursive Theory of Racial Identity: *The Souls of Black Folk* as a Response to Nineteenth-Century Biological Determinism." *Western Journal of Communication* 63, no. 2: 193–215.

Wilson, T. Anansi. "Furtive Blackness: On Blackness and Being." *Hastings Constitutional Law Quarterly* 48, no. 1 (2020): 141–79.

Wynter, Sylvia. "Unsettling the Coloniality of Being/Power/Truth/Freedom: Towards the Human, After Man, Its Overrepresentation—an Argument." *New Centennial Review* 3, no. 3 (2003): 257–337.

Yergeau, M. Remi. *Authoring Autism: On Rhetoric and Neurological Queerness*. Durham, NC: Duke University Press, 2018.

Young, Stella. "I'm Not Your Inspiration, Thank You Very Much." Filmed April 2014. TEDxSydney, video, 9.03, posted June 9, 2014. https://www.ted.com/talks/.

Yusoff, Kathryn. *A Billion Black Anthropocenes or None*. Minneapolis: University of Minnesota Press, 2018.

Index

A

F

G

N

O

P

Q

R

Y

John M. Ackerman is an associate professor in the Program for Writing and Rhetoric at University of Colorado, Boulder, and is jointly affiliated with the departments of communication and English. His research reveals the material history, representation, and performance of neighborhoods, cities, and regions in the U.S. northeast and southwest. His current research explicates "doctrines of discovery" and other settler colonial technologies that code and contain White supremacy in urban life and set the terms and conditions for antiracist and uneven economic recovery and renewal.

Chris Ingraham is an associate professor of communication, and core faculty member in Environmental Humanities at the University of Utah. His interdisciplinary teaching and research draw on rhetorical theory, environmental communication, and media aesthetics to make sense of the many environments that humans create and inhabit. He is the author of *Gestures of Concern* and coeditor, with Nicholas Taylor, of *LEGOfied: Building Blocks as Media*.

Jennifer Lin LeMesurier is an associate professor at Colgate University who studies the intersection of embodiment, culture, and race. Her scholarship has been published in *Rhetoric Society Quarterly*, *Rhetoric Review*, and *College Composition and Communication*. Her book, *Inscrutable Eating: Asian Appetites and the Rhetoric of Racial Consumption*, analyzes how rhetorics of food dovetail with the portrayal of Asians in Western contexts.

Bridie McGreavy is an associate professor of environmental communication in the Department of Communication and Journalism at the University of Maine. McGreavy is also a faculty fellow in the Senator George J. Mitchell Center for Sustainability Solutions. She studies how communication shapes sustainability and justice efforts in coastal shellfishing communities, river restoration and freshwater conservation initiatives, and diverse collaborations to address complex problems. Her work has been published in an interdisciplinary set of journals and books.

Candice S. Rai is an associate professor of English at the University of Washington. She recently coedited *Writing Across Difference: Theory and Intervention* (with James Rushing Daniel and Katie Malcolm) and *Field Rhetoric:*

Ethnography, Ecology, and Engagement in the Places of Persuasion (with Caroline Gottschalk Druschke) and is the author of *Democracy's Lot: Rhetoric, Publics, and the Places of Invention*. Her work engages in place-based inquiry to study public rhetoric and writing, political discourse and action, and argumentation.

Nathan Stormer is a professor of communication and journalism at the University of Maine. He has written about the history of abortion rhetoric in the United States and various themes within rhetorical theory.